# Urban Transformations, Land-use, and Environmental Change: Quantitative Approaches for Territorial Data

# RIVER PUBLISHERS SERIES IN CHEMICAL AND ENVIRONMENTAL ENGINEERING



The River Series on Environmental Engineering is a series of comprehensive academic and professional books which focus on Environmental Engineering subjects. The series focuses on topics ranging from theory to policy and technology to applications.

Books published in the series include research monographs, edited volumes, handbooks and textbooks. The books provide professionals, researchers, educators, and advanced students in the field with an invaluable insight into the latest research and developments.

Topics included in this series are as follows:-

- Climate Change Mitigation
- Environmental Engineering
- Environmental Systems Monitoring and Analysis
- Green and sustainable engineering techniques for modern manufacturing
- Life cycle engineering
- Sustainability in mechanical, manufacturing and industrial engineering
- Water and waste water Management

For a list of other books in this series, visit www.riverpublishers.com

# Urban Transformations, Land-use, and Environmental Change: Quantitative Approaches for Territorial Data

**Editor**

**Margherita Carlucci**
Sapienza University of Rome, Italy

River Publishers

Routledge
Taylor & Francis Group
NEW YORK AND LONDON

**Published 2023 by River Publishers**
River Publishers
Alsbjergvej 10, 9260 Gistrup, Denmark
www.riverpublishers.com

**Distributed exclusively by Routledge**
605 Third Avenue, New York, NY 10017, USA
4 Park Square, Milton Park, Abingdon, Oxon OX14 4RN

*Urban Transformations, Land-use, and Environmental Change: Quantitative Approaches for Territorial Data* / by Margherita Carlucci.



Routledge is an imprint of the Taylor & Francis Group, an informa business

ISBN 978-87-7022-738-4 (print)
ISBN 978-10-0084-653-9 (online)
ISBN 978-10-0337-325-4 (ebook master)

# Contents

# Acknowledgment

The publication of this book was funded by Sapienza University of Rome under the research project POLY-DESERT.

# List of Contributors

**Bianchini, Leonardo,** *Department of Agricultural and Forest Science, University of Tuscia, Viterbo, Italy*

**Carlucci, Margherita,** *Department of Social and Economic Sciences, Sapienza University of Rome, Rome, Italy*

**Ciaschini, Clio,** *Department of Economic and Social Sciences, Marche Polytechnic University, Ancona, Italy*

**Ciommi, Mariateresa,** *Department of Economic and Social Sciences, Marche Polytechnic University, Ancona, Italy*

**Egidi, Gianluca,** *Department of Agricultural and Forest Science, University of Tuscia, Viterbo, Italy*

**Ermini, Barbara,** *Department of Economic and Social Sciences, Marche Polytechnic University, Ancona, Italy*

**Imbrenda, Vito,** *Institute of Methodologies for Environmental Analysis, National Research Council of Italy (IMAA-CNR), 85050, Tito Scalo, Italy*

**Lamonica, Giuseppe Ricciardo,** *Department of Economic and Social Sciences, Marche Polytechnic University, Ancona, Italy*

**Mariani, Francesca,** *Department of Economic and Social Sciences, Marche Polytechnic University, Ancona, Italy*

**Polinesi, Gloria,** *Department of Economic and Social Sciences, Marche Polytechnic University, Ancona, Italy*

**Quaranta, Giovanni,** *Department of Mathematics, Computer Science and Economics, University of Basilicata, Potenza, Italy*

**Recchioni, Maria Cristina,** *Department of Economic and Social Sciences, Marche Polytechnic University, Ancona, Italy*

**Rontos, Kostas,** *Department of Sociology, University of the Aegean, Greece*

**Salvati, Luca,** *Department of Methods and Models for Economics, Territory and Finance (MEMOTEF), Sapienza University of Rome, Rome, Italy*

**Salvia, Rosanna,** *Department of Mathematics, Computer Science and Economics, University of Basilicata, Potenza, Italy*

**Salvucci, Gianluigi,** *Italian National Institute of Statistics (ISTAT), Italy*

**Samela, Caterina,** *Institute of Methodologies for Environmental Analysis, National Research Council of Italy (IMAA-CNR), 85050, Tito Scalo, Italy*

**Zitti, Marco,** *c/o Italian Council for Agricultural Research and Economics, Centre of Research in Forestry and Woods (FL), Rome, Italy*

# List of Figures

# List of Tables

# List of Abbreviations

| | |
|---|---|
| CBD | Central Business District |
| CLC | Corine Land Cover |
| CORINE | Coordination of Information on the Environment |
| ELSTAT | Hellenic Statistical Authority |
| GDP | Gross Domestic Product |
| GRMP | General Region Master Plan |
| ISPRA | Istituto Superiore per la Protezione e la Ricerca Ambientale |
| ISTAT | Istituto Nazionale di Statistica |
| NUTS | Nomenclature of Territorial Units for Statistics |
| OECD | Organization for Economic Co-operation and Development |
| PCA | Principal Component Analysis |
| PIAAC | Programme for the International Assessment of Adult Competencies |
| PRG | Master Plan |
| PUR | Polycentric Urban Region |

# 1

# Urban Transformations, Land-use, and Territorial Change Before and After Crisis: A Mediterranean Perspective

**Margherita Carlucci[1] and Luca Salvati[2]**

[1]Department of Social and Economic Sciences, Sapienza University of Rome, Rome, Italy
[2]Department of Methods and Models for Economics, Territory and Finance (MEMOTEF), Sapienza University of Rome, Rome, Italy

## 1.1 Introduction

Among the processes that have marked the socioeconomic evolution of the last 50 years in terms of intensity of effects and speed of diffusion, an extremely important role is played by the growth of the urban population, which, in Europe, accounts for more than two-thirds of the total, and the degradation of the lands, with the intensification of the processes of alteration of the climate which has translated into a growing deterioration of the quality of the fertile soil. The two phenomena mentioned above are intimately connected. Situations in which the expansion of cities has resulted in an uncontrolled diffusion of built-up surfaces in peri-urban areas, with considerable costs in terms of agricultural and natural land consumption, have favoured the increase in land degradation processes and desertification.

In the last decades, European regions have undergone a continuous growth of sparse, low-density settlements, both within the surroundings of core cities and beyond. Mediterranean countries, traditionally characterized by compact development along urban–rural gradient, have shown increasing evidence of disperse settlements progressively extending on rural districts away from metropolitan regions. The massive expansion of build-up areas has long been recognized as a major threat to sustainable land use. Urban sprawl appropriates landscape, and since soil is a finite resource, land take competes with the need of land for food production, for plants producing renewable energy, and for recreational space and niches to preserve biodiversity. Urbanization processes of the "sprawl" type, with non-contiguous residential and commercial settlements dispersed over the territory ("leapfrogging"), or developed along a road or coastal axis ("edge-expansion"), affect land quality in terms of subtraction of soils for residential buildings and transport infrastructures, but also due to the growing spread throughout the territory of landfills and often uncontrolled mining activities, which are also correlated with contamination processes that lead to further degradation. The risk of desertification involves both areas particularly suited to agriculture and regions characterized by a greater industrial presence and services. Here, in the face of a productive activity often with high profitability, there is a progressive degradation of the land, not so much linked to the processes of agricultural intensification (think of the compaction of the soils, the degradation due to the total organic load, and the pressure of grazing) as to soil erosion phenomena, due to the lack of management of the territory typical of the rural and peri-urban space, which causes landslides.

## 1.2 The Quest for Permanent Monitoring and Quantitative Assessment of Complex Socioeconomic Dynamics

Permanent monitoring of demographic and economic processes as potential drivers of sprawl is thus a crucial task for a sustainable regional development. However, a comprehensive analysis of the evolutionary dynamics of regional systems requires an in-depth investigation of (apparent and latent) spatial disparities in socioeconomic variables, under multifaceted conditions responding differently to economic expansion and recession. More specifically, this book, focused on Mediterranean regions that have undergone important changes in the economic base and in the demographic composition, aims to offer a contribution to the understanding of these complex issues. In the first part of the book, attention was paid to the construction of integrated datasets of official statistics, able to provide a comprehensive representation of the processes in place. Subsequently, studies investigating specific development paths for homogeneous groups of local districts were presented, benefiting of an explorative framework, which incorporated descriptive and inferential statistical methods.

Based on these premises, the present book is organized in six chapters contributing to the main idea of the study. Starting from urbanization, a socioeconomic process intended as particularly sensitive to economic downturns, Chapter 2 focuses on the traditional, long-established idea of inclusivity and equality of the so-called "European City", which has been recently threatened by increasingly faster urbanization processes more sensitive than ever to recession. As a particularly evident form of urbanization, urban sprawl emerged as a fragmented and unbalanced development path responding (more or less rapidly) to economic stimuli. Such dynamics are likely to produce severe consequences in the socioeconomic sphere of individuals, leading, in particular, to income inequalities. Within this context, the work provides a theoretical framework aimed at analysing the present urban expansion paths. The contribution is addressed to policy makers aimed at providing indications to design effective socio-territorial policies to enhance the quality and equity of future urbanization in metropolitan areas throughout the world.

## 1.3 Environmental Challenges, Socioeconomic Dynamics, and Sustainable Development

A serious challenge to sustainable development comes from the pressure of growing touristic demand on sensitive coastal areas that are gradually losing their rural profile in favour of an impressive diffusion of low-density settlements, with massive losses of pervious land together with increasing burden on local infrastructural carrying capacity. Chapter 3 analyses a paradigmatic example of tourism-driven processes of sprawl occurring spatially far away from (and functionally outside the direct influence of) urban regions, namely the isle of Mykonos, Greece. The relevance of this case study relies on the assessment of Mykonos land consumption due to sprawl through the integration of statistical data from multiple sources: the censuses of population, buildings, and agriculture, land-use and cadastral maps, as well as a field survey. Empirical findings illustrate, from a morphological point of view, the different phases of urban expansion in Mykonos and indicate some potential causes of this complex process. Thus, the study offers an example of integrated sprawl monitoring applied to an originally rural area with implications for environmental management and regional planning at the Mediterranean scale.

Consolidating the main idea of this study, i.e., urbanization as a socioeconomic process sensitive to economic cycles, Chapter 4 analysed population trends over 140 years (1871–2011) at municipal level in Latium, Central Italy. A multivariate approach including principal components and cluster analyses has been followed, in order to analyse the formation and consolidation of a coastal-inland gradient in population density. Results pointed out the importance of internal migrations along both elevation (mountain areas versus lowlands) and urban–rural gradients (Rome versus countryside) creating a socioeconomic gap among lowland, intensified agricultural zones, and inland mountain areas, which consolidated with the recent economic downturns. Differences in population density are further increasing due to peri-urbanization and coastalization processes. The study definitely highlights the importance of exploratory analysis in spatial population dynamics.

## 1.4 Urbanization, Resilience, and Sustainability

Urban expansion adapts to rural areas, as far as shape and structure, on the basis of the local socioeconomic backgrounds and the spatial configuration of forms. Due to urbanization pressure in rural areas, landscapes significantly transformed over time. The issue of sensitivity of socioeconomic systems to expansion and crisis clearly goes beyond the metropolitan regions, involving rural districts all over Europe with different intensity and modalities. In traditional peripherical regions of Europe, urban dispersion has become to blend the typical rural landscapes with a new residential settlement matrix based on low-density patches. For instance, "rural" sprawl has been particularly intense, thanks to the lack of compliance with residential planning rules. Chapter 5 provides a specific analysis of such dynamics focusing on Tuscia (province of Viterbo, near the city of Rome), a rural district located in Central Italy, being characterized by traditional crops of high agricultural value coexisting with urban dispersion due to "tolerant" urban plans. In these areas, sprawl was intrinsically associated with agricultural activities. This chapter postulates that, while the study area is more liveable compared to the neighbourhoods around the city of Rome, urbanization processes in Tuscia are shaped by mixed processes of socioeconomic transformations associated with rural population trends that depend on the primary sector. This, in turn, has contributed to preserve a strong naturalness of the local context. Ultimately, Tuscia is probably one of the most pertinent Mediterranean local contexts where spatial dynamics have better adapted to rural communities, influencing the long-term development growth.

Based on long-term dynamics characteristic of socioeconomic systems, the existence of a latent relationship between education level, on one side, and economic structures and value added production, on the other, has been demonstrated and investigated in depth, assuming this relationship as a distinctive trait of transitional systems responding differently to economic downturns. It is well-known that a higher level of study is associated with better levels of health, a higher probability of obtaining a job, and higher perceived incomes. For these reasons, interest in analysing academic performance has grown in recent years. Data from the Education at a Glance annual report shows that there is great disparity in academic performance in OECD countries, particularly in terms of tertiary education. It emerges that more adults between the ages of 25 and 34 are attaining tertiary degrees than adults between 55 and 64 years. Moreover, for individuals aged 30–34 years who achieve a tertiary level of education, adults with at least one tertiary-educated parent are still more likely to attain a tertiary degree than individuals from families whose parents do not have a tertiary level of education. To measure this phenomenon, two aspects must be analysed: mobility and opportunity. "Opportunity" means the probability for a child to attain tertiary education regardless of the level of education achieved by the parents. "Mobility" implies the similarity in probabilities of achieving tertiary education between children from families with a different level of education. Based on these premises, Chapter 6 introduces an index for each of these aspects and provides a broad discussion deriving from an empirical ranking of OECD countries according to these two indices. A comparison of the two rankings reveals conflicting results in some cases. Countries that occupy high positions in one ranking do not necessarily occupy equally high positions in the other. To overcome this contradiction, different aggregate methods to combine the two indices in order to provide a ranking that accounts for both dimensions are proposed.

## 1.5 Concluding Remarks

Taken together, the results obtained in the specific contributions of this book provide interesting insights for the identification of socioeconomic, demographic, and territorial factors that affect structural disparities in local economies. Morphology and intensity of urban development are, in fact, the result of demographic dynamics at the local level, connected to socioeconomic factors, and of construction activities, whose fluctuations are particularly sensitive to the economic cycle in countries, such as the ones in the Mediterranean basin, characterized by greater informality of the sector and limited public/social housing. Our objective is to provide a cognitive contribution to sustainability planning, explaining the link between forms of urban development and soil degradation and the role played in forging this relationship by the national and local economic context, in order to be able to provide policy indications for the integrated territorial planning, aimed to the containment of land degradation processes particularly in urbanized areas, thus providing an important information support to cohesion policies leveraging social and economic competitiveness.

## Acknowledgments

The publication of this book was funded by Sapienza University of Rome under the research project POLY-DESERT.

## References

Adger, W. N. (2000). Social and ecological resilience: are they related? Progress in Human Geography, 24, 347.

Ahern, J. (2013). Urban landscape sustainability and resilience: the promise and challenges of integrating ecology with urban planning and design. Landscape Ecology, 28(6), 1203-1212.

Albrechts, L., Barbanente, A., & Monno, V. (2020). Practicing transformative planning: the territory-landscape plan as a catalyst for change. City, Territory and Architecture, 7(1), 1-13.

Andersen, A. D., & Andersen, P. D. (2017). Foresighting for inclusive development. Technological Forecasting and Social Change, 119, 227-236.

Antrop, M. (2006). Sustainable landscapes: contradiction, fiction or utopia? Landscape and Urban Planning, 75(3-4), 187-197.

Baron, M. (2012). Do we need smart cities for resilience. Journal of Economics & Management, 10, 32-46.

Barthel, S., & Isendahl, C. (2013). Urban gardens, agriculture, and water management: sources of resilience for long-term food security in cities. Ecological Economics, 86, 224-234.

Beilin, R., Reichelt, N., & Sysak, T. (2015). Resilience in the transition landscapes of the peri-urban: from 'where' with 'whom' to 'what'. Urban Studies, 52(7), 1304-1320.

Bezold, C. (2010). Lessons from using scenarios for strategic foresight. Technological Forecasting and Social Change, 77(9), 1513-1518.

Biasi, R., Brunori, E., Smiraglia, D., & Salvati, L. (2015). Linking traditional tree-crop landscapes and agro-biodiversity in Central Italy using a database of typical and traditional products: a multiple risk assessment through a data mining analysis. Biodiversity and Conservation, 24(12), 3009-3031.

Briguglio, L. (2016). Exposure to external shocks and economic resilience of countries: evidence from global indicators. Journal of Economic Studies, 43(6), 1057-1078.

Bristow, G., & Wells, P. (2005). Innovative discourse for sustainable local development: a critical analysis of eco-industrialism. International Journal of Innovation and Sustainable Development, 1, 168-179.

Bristow, G., & Healy, A. (2015). Crisis response, choice and resilience: insights from complexity thinking. Cambridge Journal of Regions, Economy and Society, 8(2), 241-256.

Carlucci, M., Chelli, F. M., & Salvati, L. (2018). Toward a new cycle: short-term population dynamics, gentrification, and re-urbanization of Milan (Italy). Sustainability, 10(9), 3014.

Carpenter, S., Walker, B., Anderies, J., & Abel, N. (2001). From metaphor to measurement: resilience of what to what? Ecosystems, 4, 765-781.

Cecchini, M., Zambon, I., Pontrandolfi, A., Turco, R., Colantoni, A., Mavrakis, A., & Salvati, L. (2019). Urban sprawl and the 'olive' landscape: Sustainable land management for 'crisis' cities. GeoJournal, 84(1), 237-255.

Chelleri, L., Schuetze, T., & Salvati, L. (2015). Integrating resilience with urban sustainability in neglected neighborhoods: challenges and opportunities of transitioning to decentralized water management in Mexico City. Habitat International, 48, 122-130.

Christopherson, S., Michie, J., & Tyler, P. (2010). Regional resilience: theoretical and empirical perspectives. Cambridge Journal of Regions. Economy and Society, 3(1), 3-10.

Cinner, J. E., & Barnes, M. L. (2019). Social dimensions of resilience in social-ecological systems. One Earth, 1(1), 51-56.

Ciommi, M., Chelli, F. M., Carlucci, M., & Salvati, L. (2018). Urban growth and demographic dynamics in southern Europe: toward a new statistical approach to regional science. Sustainability, 10(8), 2765.

Corona, P. (2018). Communicating facts, findings and thinking to support evidence-based strategies and decisions. Annals of Silvicultural Research, 42, 1-2. DOI: 10.12899/ASR-1617.

Delfanti, L., Colantoni, A., Recanatesi, F., Bencardino, M., Sateriano, A., Zambon, I., & Salvati, L. (2016). Solar plants, environmental degradation and local socioeconomic contexts: a case study in a Mediterranean country. Environmental Impact Assessment Review, 61, 88-93.

Desouza, K. C., & Flanery, T. H. (2013). Designing, planning, and managing resilient cities: a conceptual framework. Cities, 35, 89-99.

Florida, R. (2010). *The Great Reset: How New Ways of Living and Working Drive Post-Crash Prosperity*. Random House Canada, Toronto.

Folke, C., Carpenter, S., Walker, B., Scheffer, M., Elmqvist, T., Gunderson, L., & Holling, C. S. (2005). Regime shifts, resilience and biodiversity in ecosystem management. Annual Review of Ecology, Evolution and Systematics, 35, 557-581.

Folke, C., Carpenter, S. R., Walker, B., Scheffer, M., Chapin, T., & Rockström, J. (2010). Resilience thinking: integrating resilience, adaptability and transformability. Ecology and Society, 15, 20.

Fratesi, U., & Perucca, G. (2018). Territorial capital and the resilience of European regions. The Annals of Regional Science, 60(2), 241-264.

Garau, C., & Pavan, V. M. (2018). Evaluating urban quality: Indicators and assessment tools for smart sustainable cities. Sustainability, 10(3), 575.

Gertler, M. S., & Wolfe, D. A. (2004). Local social knowledge management: community actors, institutions and multilevel governance in regional foresight exercises. Futures, 36(1), 45-65.

Gunderson, L. H., & Holling, C. S. (2002). *Panarchy: Understanding Transformations in Human and Natural Systems*. Island Press, Washington.

Hudson, R. (2009). Resilient regions in a uncertain world: wishful thinking or a practical reality. Cambridge Journal of Regions, Economy and Society, 3, 11-25.

Janin Rivolin, U. (2017). Global crisis and the systems of spatial governance and planning: a European comparison. European Planning Studies, 25(6), 994-1012.

Kaur, H., & Garg, P. (2019). Urban sustainability assessment tools: a review. Journal of Cleaner Production, 210, 146-158.

Knieling, J., & Othengrafen, F. (2015). *Cities in Crisis: Socio-Spatial Impacts of the Economic Crisis in Southern European Cities*. Routledge, London.

Komninos, N., Kakderi, C., Panori, A., & Tsarchopoulos, P. (2019). Smart city planning from an evolutionary perspective. Journal of Urban Technology, 26(2), 3.

Martin, R., & Gardiner, B. (2019). The resilience of cities to economic shocks: a tale of four recessions (and the challenge of Brexit). Papers in Regional Science, 98(4), 1801-1832.

Martini, B. (2020). Resilience and economic structure. Are they related? Structural Change and Economic Dynamics, 54(C), 62-91.

Medeiros, E. (2017). From smart growth to European spatial planning: a new paradigm for EU cohesion policy post-2020. European Planning Studies, 25(10), 1856-1875.

Meerow, S., Newell, J. P., & Stults, M. (2016). Defining urban resilience: a review. Landscape and Urban Planning, 147, 38-49.

Moulaert, F., Martinelli, F., Swyngedouw, E., & Gonzalez, S. (2005). Towards alternative model (s) of local innovation. Urban Studies, 42(11), 1969-1990.

Nadin, V., Stead, D., Dąbrowski, M., & Fernandez-Maldonado, A. M. (2021). Integrated, adaptive and participatory spatial planning: trends across Europe. Regional Studies, 55(5), 791-803.

Olsson, E. G. A., Kerselaers, E., Søderkvist Kristensen, L., Primdahl, J., Rogge, E., & Wästfelt, A. (2016). Peri-urban food production and its relation to urban resilience. Sustainability, 8(12), 1340.

Parr, J. B. (2007). Spatial definitions of the city: four perspectives. Urban Studies, 44(2), 381-392.

Percoco, M. (2017). Impact of European Cohesion Policy on regional growth: does local economic structure matter? Regional Studies, 51(6), 833-843.

Perrin, C., Nougarèdes, B., Sini, L., Branduini, P., & Salvati, L. (2018). Governance changes in peri-urban farmland protection following decentralisation: a comparison between Montpellier (France) and Rome (Italy). Land Use Policy, 70, 535-546.

Recanatesi, F., Clemente, M., Grigoriadis, E., Ranalli, F., Zitti, M., & Salvati, L. (2016). A fifty-year sustainability assessment of Italian agro-forest districts. Sustainability, 8(1), 32.

Repo, P., & Matschoss, K. (2020). Social innovation for sustainability challenges. Sustainability, 12(1), 319.

Salvati, L., & Zitti, M. (2007). Territorial disparities, natural resource distribution, and land degradation: a case study in southern Europe. Geojournal, 70(2), 185-194.

Salvati, L., Gemmiti, R., & Perini, L. (2012). Land degradation in Mediterranean urban areas: an unexplored link with planning? Area, 44(3), 317-325.

Sánchez, M. L., Cabrera, A. T., & Del Pulgar, M. L. G. (2020). Guidelines from the heritage field for the integration of landscape and heritage planning: a systematic literature review. Landscape and Urban Planning, 204, 103931.

Simmie, J., & Martin, R. (2010). The economic resilience of regions: towards an evolutionary approach. Cambridge Journal of Regions, Economy and Society, 3, 27-43.

Stobbelaar, D. J., & Pedroli, B. (2011). Perspectives on landscape identity: a conceptual challenge. Landscape Research, 36(3), 321-339.

Tsiapa, M., Kallioras, D., & Tzeremes, N. G. (2018). The role of path-dependence in the resilience of EU regions. European Planning Studies, 26(6), 1099-1120.

Van der Sluis, T., Pedroli, B., Frederiksen, P., Kristensen, S. B., Busck, A. G., Pavlis, V., & Cosor, G. L. (2019). The impact of European landscape transitions on the provision of landscape services: an explorative study using six cases of rural land change. Landscape Ecology, 34(2), 307-323.

Walker, B., Gunderson, L., Kinzig, A., Folke, C., Carpenter, S., & Schult, L. (2006). A handful of heuristics and some propositions for understanding resilience in social- ecological systems. Ecology and Society 11, 13.

Wilson, B., & Chakraborty, A. (2019). Planning smart(er) cities: The promise of civic technology. Journal of Urban Technology, 26(4), 29-51.

Yigitcanlar, T., Kamruzzaman, M., Foth, M., Sabatini-Marques, J., da Costa, E., & Ioppolo, G. (2019). Can cities become smart without being sustainable? A systematic review of the literature. Sustainable Cities and Society, 45, 348-365.

Zambon, I., Benedetti, A., Ferrara, C., & Salvati, L. (2018). Soil matters? A multivariate analysis of socioeconomic constraints to urban expansion in Mediterranean Europe. Ecological Economics, 146, 173-183.

# 2

# The Future Paths of Urbanization: A Brief Review

**Clio Ciaschini and Barbara Ermini**

Department of Economic and Social Sciences, Marche Polytechnic University, Ancona, Italy

## Abstract

Nowadays, the traditional, established idea of inclusivity and equality of the so called "European City" has been threatened by an increasingly faster urbanization process. The present urban sprawl follows a fragmented and unbalanced development path. This fragmentation is likely to produce severe consequences in the socioeconomic sphere of individuals, leading, in particular, to income inequalities. Within this context, the work provides a theoretical framework aimed at analysing the present urban expansion paths. The contribution is addressed to policy makers aimed at providing indications to design effective socio-territorial policies to enhance the quality and equity of future urbanization in metropolitan areas throughout the world.

## 2.1 Introduction

In recent decades, continuous urbanization has been influenced by the joint interaction of population growth and migration (Bettencourt *et al.*, 2007; Chen *et al.*, 2014). The present globalization era has not weakened the relevance of large "global city regions", the driving force of the global economy. Hall and Pain (2006) defined polycentric city regions as "a series of anything between ten and fifteen cities and towns, physically separated but functionally networked, clustered around one or more larger central cities, and drawing enormous economic strength from a new functional division of labor. [...] It is no exaggeration to say that this is the emerging urban form at the start of the 21st century". Polycentric spatial organizations often reflected a dispersed and discontinuous urban form (Jonas and Ward, 2007) with a high rate of land consumption (Salvati, 2014). Throughout the years, the "global" cities acquire the ability to reorient the national and international economy (Scott, 2019) affecting both social and cultural systems (Hall, 2009). Global cities, indeed, include several sectors and productions (Celant, 2007). Moreover, with reference to firm attractiveness and economic potential, these cities undertook relevant transformations toward polycentrism (Davoudi, 2008; McCann, 2007; Rodriguez-Pose, 2008). The economic costs of urban form are characterized by three main elements, i.e., density, centricity, and city size. With reference to density, both developed and developing countries seem to tend to a de-densification of urban areas, increasing decentralization and fragmentation, together with growth of both population and land area (Salvati, 2015). This phenomenon of de-densification of urban areas implies, among others, higher transportation costs and higher greenhouse per capita gas emissions together with a decreasing productivity. High-density city levels imply congestion and high land prices; despite this fact, in these areas, the produced benefits exceed the relevant incumbent costs particularly in less developed countries (Rauch, 1991). The present combination of decentralization–de-densification seems to be inefficient even if empirical studies highlight that polycentric structures minimize costs due to high density and large population. In general, the variation of city size is strictly linked to the effects of density and centricity. Higher costs derived from the shift of citizens from high-density urban sites to low density cities induce policy makers to design policy measures to address the relevant issue of de-densification, despite the resistance of government and the durability of fixed capital investments (Kim and Rowe, 2012). Large city size, generating economies of agglomeration, usually imply higher wages, higher proportions of educated citizens, and higher productivity, also in correspondence of augmented proximity and scale of large cities. The main benefits of economies of agglomeration are proximity and scale, which facilitate knowledge and human capital spill-overs and reduce costs of investments and transactions (Brugmann, 2012; Segal, 1976). Within these economies, advantages can also result in firm's access to wider markets. These benefits are largely compensated by higher costs of living, congestion, and increasing inequalities (Duranton *et al.*, 2009). Benefits of agglomeration include both benefits of increased city sizes and densities, implying greater chances for productive improvement and knowledge spill-overs and shared infrastructures (lower fixed costs). Precisely speaking, multicentricity defines all urban regions that contain multiple centres, independently from their size and importance, while polycentrism, generally known as polycentric urban region (PUR), describes a particular case of multicentric regions with all centres having similar size and importance (Kloosterman and Musterd, 2001; Parr, 2012). The background of literature counts not many studies dealing with the formation of PUR. The most relevant analyses have been performed by Champion (2001), and lately by Shu *et al.* (2019), that designed how functional urban regions could rise on the course of time, but these studies have not still found any practical application. In addition, not many existing studies on polycentricity consider the dynamics of population, even if the demographic aspect is fundamental in the definition and identification of a PUR from other forms of urban regions. Meijers *et al.* (2018), indeed, defined 117 European PURs following the maximum variation of differences in population sizes between the cities of a PUR. A relevant contribution in the demographic analysis is provided by the population change that, more than the absolute population numbers, is informative on the urban transformation, over a long-time horizon. This indicator constitutes both an influence and a consequence of socioeconomic and spatial development processes (Turok and Mykhnenko, 2007), easily comparable across places and time periods.

## 2.2 Monocentricity vs. Polycentricity

One of the most important theoretical frameworks adopted in the research on the determinants of urban form is that based on the Alonso–Mills–Muth (AMM) monocentric model of urban spatial structure. The basic hypothesis of this static model is the existence of a city with a stated level of population and a fixed income level, living in the vicinity of a central business district (CBD) (Takahashi, 2014). Each worker must reach the city centre to work. At stated fixed conditions, workers are prone to live closer to the city centre, also due to the high costs of commuting and due to the fact that these costs increase in proportion of the distance from the city centre. Despite this intention, not all workers can live in the vicinity of the centre; this implies an adjustment in prices and density of housing to equilibrate the market. Particularly speaking, in the vicinity of the CBD, land for housing is more expensive; homebuilders are so induced to build multi-storeyed buildings and smaller dwellings since the cost of an additional square metre increases with building height. The two possible decisions for the households are: (i) living in smaller and more expensive housing in the city or (ii) in cheaper structures less close to the city centre (Brueckner and Fansler, 1983). The determinants of the size of a city are mainly three: the size of the population, the cost of transport, and the value of land in uses alternative to housing (like agriculture). In the model formalization, all the citizens are identical with reference to both income and preference. Households' preferences in housing are formalized as their demand for generic housing; extension of this hypothesis is done by including the demand for yard space, but conclusions appeared to be the same (Arnott and MacKinnon, 1977). The model excludes the polycentric city or locations desirable for education or leisure facilities; only the assumption of monocentric city is allowed. Also in this case, results appear to be robust with reference to models of the "polycentric" type (Spivey, 2008). In addition, also wage rate, the labour, and capital market are taken as exogenous. Despite its simple and rather unrealistic structure, it succeeds in the purpose of explaining the impact of the different policies and demographic growth on the urban equilibrium (Kulish *et al.*, 2012). Even if empirical evidence exists, which proves that monocentric theory succeeds in modelling robustly urban forms (Paulsen, 2012), the prevailing theoretical wave tends to shift from the monocentric framework to a polycentric one (Durbin and Sung, 1987). In developed and developing urban areas, in fact, de-densification and increasing decentralization and fragmentation spread together with increase in city size with reference to both population and land area (Anderson *et al.* 1996; Angel *et al.*, 2011; Istrate *et al.*, 2011). The main long-term effects on the form of the city are determined by the policy environment in which an area urbanizes. In this way, different policy regimes generate cities with largely diverse profiles (Bertaud and Malpezzi, 2003) and different prices to control city externalities influence the behaviour of individuals in producing more efficient city forms (Brueckner *et al.*, 2001; Gordon and Richardson, 1996). Effective urban policies are, actually, difficult to implement also because their realization is subordinate to a large amount of fixed capital investments, human capitals, and political issues. For decades, the monocentric urban model based on single CBD prevailed in the urban analysis. In more recent years, the key assumptions of this approach have been criticized from different point of views. In particular, scholars dealing with public policy challenged the undiscussed hegemony of monocentricity: most of the cities, in fact, converged into structures with different employment centres, raising the topic of contrast between the simple monocentric theory and the more complex reality (Arribas-Bel and Sanz-Gracia, 2014). Within this framework, Fujita and Ogawa (1982) designed a theoretical model of urban polycentricity, with the possible outcomes of multiple equilibria. The main hypothesis of this framework is the modelization of the benefits from interactions between firms by means of a negative exponential function of distance (differently from the usual linear relation). Contrarily to the traditional monocentric models, present polycentric models presume endogenously determined employment centres on the basis of the agents' behaviour and urban environment (White, 1999). This theory assumes that, above a certain threshold, modern metropolitan areas are defined as polycentric (this category includes the majority of the existing metropolitan areas). The debate on definition of polycentricity and polycentric urban regions (PUR) is still open within scholars dealing with urban studies. A unanimous consensus has been reached only on a basic definition, i.e., a polycentric region is a spatially contiguous area with different (poly, more than one) centres located between a nation, state, and an individual city or region (Burger and Meijers, 2016; Van Meeteren, 2016). In addition, in a polycentric system, a balanced distribution of demographic and economic growth is considered a suitable midpoint between excessive density and urban sprawl (WBGU, 2016). At a regional scale, e.g., in European context, the concept of PURs is defined in different ways: "polycentric urban region" (Kloosterman and Musterd, 2001; Meijers *et al.*, 2018), "global city-region" (Scott, 2019), "mega-city region" (Hall and Pain, 2006), or "polycentric metropolitan region". The path of formation of a PUR undergoes to two main lines, i.e., "intra-urban" polycentricity, i.e., polycentric spatial structures arisen in post-industrial or post-suburban developments in the hinterlands of traditional monocentric core cities, and "inter-urban" polycentricity that develops in functional interconnections of spatially close core cities and the augmenting overlap of cities with their respective suburbs (Kloosterman and Musterd, 2001). Examples of inter-urban regions are provided by the Rhine-Ruhr region (Germany), the Randstad (the Netherlands), and the Upper Silesian industrial region, (Poland).

## 2.3 Demographic Dynamics in Metropolitan Areas

Worldwide, between early 1970s and 1990s, metropolitan clusters grew at relevant pace, in correspondence of demographic changes and increases in population (Fielding, 1982; Champion, 1989; Cross, 1990; Fournier *et al.*, 2020). Different methodologies and approaches have been implemented to deal with the issue of population redistribution over expanding metropolitan regions, such as indicators and spatial scales of analysis (Lever, 1993; Mordridge and Parr, 1997; Van Criekingen, 2010). Starting from 1990, a transition to more spatially complex urban dynamics, with socioeconomic peculiarities and heterogeneous demographic processes, is observed in

Mediterranean cities (Carlucci *et al.*, 2017). The most relevant variable for the study of socioeconomic transformations is the long-term population (Wilson, 2001; Zitti *et al.*, 2015).

A metropolitan life cycle is defined as a time interval with rise and decrease of a given urban phase associated with some demographic dynamics, social trends, and economic performances (Forster, 2006; Davis, 2016). The metropolitan development in a certain country has been effectively studied by means of the analysis of urban growth by size class of towns (Rogers, 1982; Seto *et al.*, 2011). The increasing relevance of the urban regions in a globalized economy implies that bigger cities grow faster than smaller ones, especially in the early phase of urbanization (Jones, 2002). In the latest phase, in most cases, small towns perform a faster growth due to congestion and crowding in large and intermediate towns (Martinez-Fernandez *et al.*, 2012). In any case, irregular behaviour in the relationship between population growth rate and initial sizes of cities exists, following the Gilbrat's law (Weber and Sciubba, 2018). Works in favour of Gilbrat's law suggested the relationship between population growth rate and city size. Particularly speaking, they show that larger cities attract and preserve political and economic hegemony within national, continental, or world urban hierarchy. Nevertheless, studies against the Gilbrat's law remark negative relationships between the size of cities and population growth rate, under the hypothesis of presence of scale effects, i.e., higher variance in the growth rate in the biggest cities than in the smallest ones, implying that smaller cities could grow faster than largest cities (González-Val, 2014; McCann, 2017). Generally, Gilbrat's framework does not succeed in explaining specific phenomena, such as "urbanization without economic growth" often detected in developing regions and, also, even if rarely, in some developed countries. Assuming that the metropolitan expansion is driven by population growth, demographic dimension of the urban growth assumes a more relevant role also in connection to more general processes, in particular the second demographic transition (Reher, 2011). The creation of the Spatial Cycle Model dates to the mid-20th century. The model of Van Den Berg *et al.* (1982) provides a widely accepted path of the population development of urban core and ring in a functional urban region. The key idea of this model is that demographic growth is different between the core and the ring, both under in timing and intensity, and it develops in four stages: urbanization, suburbanization, dis-urbanization/counter-urbanization, and re-urbanization. In the phase of urbanization, the growth is related primarily to the core and then to the ring areas. The suburbanization witnesses a population drop in the core and an increase in the ring area. Dis-urbanization presents a decrease in the core and the ring. The concluding phase, i.e., re-urbanization, features a re-growth of the core and a sluggish decrease in the ring. This final step can lead to the restart of the cycle from step I. The four stages can perform both balanced and counter-cyclical phases and the growth of core versus rings can also decline. Criticism on the restart of the cycle after step IV has been raised by several scholars (Egidi *et al.*, 2020; Nyström, 1992; Di Feliciantonio *et al.*, 2018). Parr (2012) provides an alternative version which confirms the four stages but is rooted on relative population change rates between core and rings, thus introducing stages of decentralization and concentration within the FUR. This new framework encourages a spatially systemic view on demographic urban-regional development since it allows the comparison of several FURs also in terms of development rates, independently from their size. This vision encompasses the permanence of a city in each stage, to move back or skip over other stages. Parr's theory is included, in any case, in the monocentric theory (urban, core, and ring). With reference to the polycentric framework, Champion (2001) considers qualitative demographic changes, with reference to population and household structures. His work compares two main trends: (1) demographic structural changes (Western Europe) to reach higher level of life expectancy and lower degrees of fertility rates (and ageing society), and augmenting volume and increasing frequency of migration flows (more mobile and varied society, defined by Van de Kaa (1987), as "second demographic transition"); (2) regional structural changes that substitute the traditional monocentric urban structures with new polycentric theories. Following Champion (2001), there are three models of urban region formation: (i) the centrifugal mode, (ii) the incorporation mode, and (iii) the fusion mode. The centrifugal mode includes the regional features of urban sprawl (Gordon and Richardson, 1996) where a multicentric urban region spreads into the hinterland, generating different sub-centres. In the incorporation mode, a wide urban centre spreads to include smaller centres located in the nearby areas independent under the point of view of growth in employment, services, and demographic trends. The growth of these large cities seems to be less influenced by the decentralization of the core, if compared with urban regions concentrated around a predominant city (Cardoso, 2018). The third mode is the so-called post-suburbia (Kling *et al.*, 1995). The hinterlands of independent and similar sized cities merge into a PUR. Shu *et al.* (2019) describe four ways of polycentralization, i.e., topographical and topological understanding, and decentralization and coalescence paths. In the topographical understanding, the characterizing features of the centre (services, jobs, etc.) are crucial, while the most relevant elements of topological understanding are the flows (in firm networks). The issue of decentralization is addressed to the relevance of sub-centres in a city, and coalescence defines the transition of originally standing alone cities that become connected. The previously discussed cyclic development of urbanization, suburbanization, dis-urbanization, and re-urbanization is completed when the society reaches an advanced polycentric structure, as in the Netherlands, where the various urban areas relate with different neighbours overcoming the distinction core-ring. Nevertheless, the status of hinterland of multiple cities has several effects in particular with reference to demographic growth (Phelps and Wu, 2011). These newly constituted areas, i.e., poly-hinterlands, are difficult to delimitate since it is difficult to define to what extent these areas could be defined as hinterlands and centre of other cities. In consideration of elements as proximity to multiple cores, that host different amenities and activities, but not the complete urban activity, the poly-hinterlands are located within the urban region and defined as central when situated in a position easily reachable by the majority of urban activities and that attracts wide flows of population. The post-modern framework, though, does not suggest a final stage of balanced equilibrium; contrarily, in some cases, cores and rings synchronize their demographic trajectories and lead to the emergence of more flexible and unpredictable intra-regional centralities and flows. Following Humer *et al.* (2021), this new stage of synchro-urbanization defines a period of urban-regional development leading to a better explanation of the polycentricity through the study of the links between core and peripheries. This new urbanization path indicates a synchronous growth and decline of different cores and regional hinterlands that will reorganize regional centralities exposing each urban region to urban systemic development across larger areas (Rauhut and Humer, 2020; Zimmermann and Atkinson, 2021). The advantage of this new urban trajectory raises debates mainly focused on the need of implementation of policies supporting sustainable urban development, in particular with reference to land consumption and ecological externalities of transport (Marquard *et al.*, 2020; Meijers and Sandberg, 2021).

## 2.4 Urban Transformations, Socioeconomic Wellbeing, and Natural Amenities

Socioeconomic development and the process of urban sprawl around the world have brought profound changes to urban landscapes and land use (Chowdhury *et al.*, 2011; Jedwab *et al.*, 2021). The territorial distribution of settlements, of both private and public nature, is more articulated in comparison to what traditional hierarchical and functional models would predict (Rickards *et al.*, 2016). The antinomy between core-periphery and vocational locations for production–consumption within a city has become less marked (Kim, 2020; Knaus *et al.*, 2021). The consequent re-organization of the territory assumed the shape of an enlarged metropolitan area filled by dispersed and fragmented developed spaces and patches, especially in the fringe and at progressively larger distances from central cities (Couch *et al.*, 2007; Arribas-Bel *et al.*, 2011; Shrestha *et al.*, 2012; Gounaridis *et al.*, 2018). It involved the redesigning of existing gradients, primarily the urban–rural, at a larger territorial level being cities' edges increasingly difficult to delimit and it also altered land use of interstitial spaces at local level creating hybrid and mixed urban and suburban landscapes (Sayadi *et al.*, 2009; Elias *et al.*, 2012; Yang *et al.*, 2018; Jia *et al.*, 2021).

The urban transformation has primarily involved the redistribution of agricultural and natural land areas (Zhou and Wang, 2011; Nor *et al.* 2017; Koprowska *et al.*, 2020). It is clear that urban green spaces provide numerous health, social, and ecological benefits. Green spaces are often associated with a wide range of ecosystem services that largely improve the maintenance and the quality of life by, for example, providing food and water, regulating services that affect climate resilience, floods, disease, wastes and water quality, reducing pollution and environmental deterioration, and supporting functions aimed at preventing soil's impoverishment (MEA 2005a, 2005b; Haase *et al.*, 2014; Leal Filho *et al.*, 2020). Under an urbanization viewpoint, stressing the cultural services provided by nature, green space is understood as natural amenities that encourage residential development given that citizens benefit, both physically and psychologically, from having access to better recreational, intellectual, aesthetic, and emotional opportunities as well as to improved environmental quality and urban biodiversity (Brueckner *et al.*, 1999; Cho, 2001; Parker and Simpson, 2018; Reisig *et al.*, 2021). Thus, agricultural and natural land areas largely affect sustainability and city liveability (García-Nieto *et al.*, 2018; Liu *et al.*, 2018; Yuan *et al.*, 2018; Ye and Becker, 2018; Heikinheimo *et al.*, 2020).

In the empirical literature, there is a large consensus that natural amenities and environmental benefits are normal goods, when not luxury (Menke, 1987; Deller *et al.*, 2001; Rappaport, 2008; Albouy *et al.*, 2020). In other words, the income elasticity of demand for environmental attributes is greater than zero, or, as income grows, environmental concern rises as well as demand for environmental goods (Grossman and Krueger, 1996; Stephens and Partridge, 2015; Carlino and Saiz, 2019; Caplan *et al.*, 2021). On this basis, the environmental Kuznets curve (ECK) framework predicts that environment quality will initially deteriorate, but some threshold of income will eventually be reached beyond which the environmental degradation will decline according to citizens demanding and affording a better environment (Chen and Wang, 2013; Isik *et al.*, 2019; Liu *et al.*, 2020). This conceptualization can be easily borrowed to interpret the preference for environmental landscape in a context of city development and urban dynamics. Urbanization is characterized by a rapid land taking and evolution of land use change, altering the long-established relationship between urban, agricultural, and natural uses of land in a sort of overwhelming commodification of the residential landscape to reflect inhabitants' preferences and lifestyles (Ströbele and Hunziker, 2017). Thus, demand of green spaces, ecological services, and natural amenities in a developing city can be driven by general economic development and by citizen's socioeconomic conditions which, however, are often characterized by highly disparity and uneven distribution over space (Burgalassi and Luzzati, 2015). It follows that the pattern of city development is strictly interconnected with the socioeconomic structure of their inhabitants (Clark *et al.*, 2002; Dalmazzo and de Blasio, 2011; Lawton *et al.*, 2013; Frenkel *et al.*, 2013; Albouy and Stuart, 2020). Socioeconomic inequality translates into observable spatial inequalities with unequal distribution of environmental amenities and landscape development raising an increasing concern about residential segregation, gentrification, and polarization in "gated communities" (Le Goix, 2005; Genis, 2007; Csefalvay and Webster, 2012; Easton *et al.*, 2020; Kostenwein, 2021) along with jurisdictional and territorial fragmentation and scattered urban development (Mingione, 2009; Cassiers and Kesteloot, 2012; Hadjimichalis, 2017; Saganeiti *et al.*, 2018).

In fact, the evolution of modern urbanization in relation to the distribution of territorial natural amenities is receiving great interest among urban planners and, above all, policy makers since they are strictly correlated to the emergence of urban segregation within metropolitan areas which tends to take a dispersed spatial form (Wu, 2006). Literature on environmental justice (Mohai and Saha, 2006; Heynen *et al.*, 2006; Downey and Hawkins, 2008; Pham *et al.*, 2012; Schwarz *et al.*, 2018) highlights that those disparities in residential landscape distribution and access to urban green amenities proved to be strictly correlated with income, education, and ethno-cultural differences (Apparicio *et al.* 2017; Wüstemann *et al.*, 2017; Anguelovski *et al.*, 2018; Vergou, 2019; Venter *et al.*, 2020). Yet, the benefits of natural amenities seem to be unevenly distributed with disadvantages borne by low income citizens and minorities groups, often culminating into urban residential segregation and gentrification with displacement of discriminated citizens (Freeman, 2009; Brueckner and Rosenthal, 2009; Black and Richards, 2020; Trounstine, 2020).

Mediterranean urban regions, for their location, i.e., in-between north-western affluent cities and developing agglomerations of the world "south", could not be fully considered as global city regions. Their characteristics, i.e., hyper-compact forms, dense settlements, poorly organized public services, and limited infrastructural networks, differ from those of the traditional global city regions (Krumholz, 1992) and act as a serious obstacle to the aim of achieving a relevant position among the richest European cities. Recent transformations in Mediterranean cities are determined by the diffusion of sparse, low-density settlements due to population de-concentration that include both rural and urban areas (Salvati and Gargiulo Morelli, 2014; Salvati and Sabbi, 2011). Nevertheless, these regions succeed in the purpose of understanding long-term dynamics of discontinuous and dispersed urban expansion for three main reasons. The first reason is linked to the urban growth policies moved by deregulation and informality (Giannakourou, 2005). The second is homogeneous demographic dynamics over space and time described by separate phasis of fast and slow population increase (Salvati, Zitti, and Sateriano, 2013). Third, cities' expansion is characterized by a complex system of interaction agents with local peculiarities, independently from land prices and zoning processes (Salvati and Gargiulo Morelli, 2014).

Understanding the diffusion of these problematic patterns of urban development is of particular relevance to monitor land evolution especially in the Mediterranean countries. Patterns of fragmented and dispersed land development, and possibly impacted by gentrification, are typical of modern urbanization processes in Northern European and oversea countries, but, in recent decades, also cities in the south of Europe began to sprawl even if they were traditionally characterized by compactness (Catalàn *et al.*, 2004; Munafò *et al.*, 2010). In addition, Mediterranean countries are witnessing practices of development of public and residential spaces attributable to urban informality so that spontaneous and unplanned activities in public areas are likely to reinforce gentrification of urban settlements (Silva, 2020). Moreover, soil erosion and degradation are serious issues in Mediterranean area territories and territorial maintenance; so protection can take place also by means of sustainable urban green land management that progresses hand in hand with the economic development and urbanization (Petanidou *et al.*, 2008; La Rosa, 2014). Nevertheless, studies focusing on competing and discriminating use of urban, agricultural, and natural land in southern European metropolitan regions are scanty, and the present paper attempts to contribute to this topic.

Evaluating the impact of city development and different urbanization modes on land use and changes may benefit from an econometric model that incorporates socioeconomic and policy factors, with analyses extending over a long time period. The correlation between urbanization-driven changes in land-use structure and average per-capita declared income by urban district can be further investigated to the aim of verifying whether recent urbanization reflected social segregation and increasing territorial disparities (Le Goix, 2005). This process determined a polarization in "gated communities" (residential districts inhabited by high-class society and surrounded by natural amenities and more fragile areas from the socio-environmental perspective (Wu, 2006).

## 2.5 Concluding Remarks

Rapid urbanization is one of the crucial issues of global change and, more and more often, we observe a remarkable correlation between socioeconomic changes and urban land use and landscape modifications (Deng *et al.*, 2009; Kabisch *et al.*, 2016). Urban expansion pattern is becoming increasingly dispersed and fragmented. Topography and natural amenities seem to contribute in shaping the territorial development in suburban districts – in particular, the expansion of urban settlement into high-quality landscapes – which, in turn, are strictly associated with preference for environmental landscape expressed by citizens characterized by uneven socioeconomic status, primarily income inequalities (Gounaridis *et al.*, 2018; Kefalas *et al.*, 2019). Possibly, the resulting socio-spatial polarization may represent a risk in terms of sustainability and liveability of modern urban area and requires a punctual spatial planning activity to restore the established traditional image of inclusivity and equality of the "European city" (Musterd *et al.*, 2017). This article discusses solid and flexible notions to evaluate urban expansion modes that are essential in envisaging well informed socio-territorial policies to contrast unbalanced trajectories of regional development. Admitting that this trend of land development is becoming a common feature of a growing number of cities in advanced and developed countries, providing instruments of evaluation of the ongoing situation, can be regarded as a useful and scientific support in the decision-making process for improving the quality and equity of future urbanization in metropolitan areas around the world.

## References

Albouy, D., Christensen, P., & Sarmiento-Barbieri, I. (2020). Unlocking amenities: Estimating public good complementarity. *Journal of Public Economics*, *182*, 104110.

Albouy, D., & Stuart, B. A. (2020). Urban population and amenities: The neoclassical model of location. *International Economic Review, 61*(1), 127-158.

Anderson, W. P., Kanaroglou, P. S., & Miller, E. J. (1996). Urban form, energy and the environment: A review of issues, evidence and policy. *Urban studies*, *33*(1), 7-35.

Angel, S., Parent, J., Civco, D. L., Blei, A., & Potere, D. (2011). The dimensions of global urban expansion: Estimates and projections for all countries, 2000–2050. *Progress in Planning*, *75*(2), 53-107.

Anguelovski, I., Connolly, J., & Brand, A. L. (2018). From landscapes of utopia to the margins of the green urban life: For whom is the new green city? *City*, *22*(3), 417-436.

Apparicio, P., Landry, S., & Lewnard, J. (2017). Disentangling the effects of urban form and socio-demographic context on street tree cover: A multi-level analysis from Montréal. *Landscape and Urban Planning*, *157*, 422-433.

Arnott, R. J., & MacKinnon, J. G. (1977). Measuring the costs of height restrictions with a general equilibrium model. *Regional Science and Urban Economics*, *7*(4), 359-375.

Arribas-Bel, D., & Sanz-Gracia, F. (2014). The validity of the monocentric city model in a polycentric age: US metropolitan areas in 1990, 2000 and 2010. *Urban Geography*, *35*(7), 980-997.

Arribas-Bel, D., Nijkamp, P., & Scholten, H. (2011). Multidimensional urban sprawl in Europe: A self-organizing map approach. *Computers, Environment and Urban Systems*, *35*(4), 263-275.

Bertaud, A., & Malpezzi, S. (2003). The spatial distribution of population in 48 world cities: Implications for economies in transition. *Center for Urban Land Economics Research, University of Wisconsin*, *32*(1), 54-55.

Bettencourt, L., M., Lobo, J., Helbing, D., Kühnert, C., & West, G. B. (2007). Growth, innovation, scaling, and the pace of life in cities. *Proceedings of the National. Academy of Science.* USA, *104*, 7301-7306.

Black, K. J., & Richards, M. (2020). Eco-gentrification and who benefits from urban green amenities: NYC's high line. *Landscape and Urban Planning*, *204*, 103900.

Brueckner, J. K., Mills, E., & Kremer, M. (2001). Urban sprawl: Lessons from urban economics [with comments]. *Brookings-Wharton Papers on Urban Affairs*, 65-97.

Brueckner, J. K., & Rosenthal, S. S. (2009). Gentrification and neighborhood housing cycles: Will America's future downtowns be rich? *Review of Economics and Statistics*, *91*, 725-743.

Brueckner, J. K., Thisse, J. F., & Zenou, Y. (1999). Why is central Paris rich and downtown Detroit poor? An amenity-based theory. *European Economic Review*, *43*, 91-107.

Brueckner, J. K., & Fansler, D. A. (1983). The economics of urban sprawl: Theory and evidence on the spatial sizes of cities. *The Review of Economics and Statistics*, 479-482.

Brugmann, J. (2012). Financing the resilient city. *Environment and Urbanization*, *24*(1), 215-232.

Burgalassi, D., & Luzzati, T. (2015). Urban spatial structure and environmental emissions: A survey of the literature and some empirical evidence for Italian NUTS 3 regions. *Cities*, *49*, 134-148.

Burger, M., & Meijers, E. (2016). Agglomerations and the rise of urban network externalities. *Papers in Regional Science*, *95*(1), 5-15.

Caplan, A. J., Akhundjanov, S. B., & Toll, K. (2021). Measuring heterogeneous preferences for residential amenities. *Regional Science and Urban Economics*, *87*, 103646.

Cardoso, R. V. (2018). Localising urbanisation trajectories: comparing the emergence of second-tier urban regions in Europe (1890–2011). *Urban Research & Practice*, *11*(3), 223-246.

Carlino, G. A., & Saiz, A. (2019). Beautiful city: Leisure amenities and urban growth. *Journal of Regional Science*, *59*(3), 369-408.

Carlucci, M., Grigoriadis, E., Rontos, K., & Salvati, L. (2017). Revisiting a hegemonic concept: Long-term 'Mediterranean urbanization' in between city re-polarization and metropolitan decline. *Applied Spatial Analysis and Policy*, *10*(3), 347-362.

Cassiers, T., & Kesteloot, C. (2012). Socio-spatial inequalities and social cohesion in European cities. *Urban Studies*, *49*(9), 1909-1924.

Catalàn, B., Sauri, D., & Serra, P. (2004). Urban sprawl in the Mediterranean? Patterns of growth and change in the Barcelona Metropolitan Region 1993-2000. *Landscape and Urban Planning*, *85*, 174-184.

Celant, A. (2007). *Global Tourism and Regional Competitiveness.* Pàtron, Bologna.

Champion, A. (2001). A changing demographic regime and evolving polycentric urban regions: Consequences for the size, composition and distribution of city populations. *Urban Studies*, *38*(4), 657-677. https://doi.org/10.1080/00420980120035277

Champion, A. (1989). Counterurbanisation: The conceptual and methodological challenge. In A. Champion (Ed.), *Counterurbanisation: The Changing Pace and Nature of Population Deconcentration*, (pp. 19–33). Edward Arnold, London.

Chen, M., Zhang, H., Liu, W., & Zhang, W. (2014). The global pattern of urbanization and economic growth: Evidence from the last three decades. *PLoS ONE*, *9*, e103799.

Chen, W. Y., & Wang, D. T. (2013). Economic development and natural amenity: An econometric analysis of urban green spaces in China. *Urban Forestry & Urban Greening*, *12*(4), 435-442.

Cho, C. J. (2001). Amenities and urban residential structure: An amenity-embedded model of residential choice. *Papers in Regional Sciences*, *80*, 483–498.

Chowdhury, R. R., Larson, K., Grove, M., Polsky, C., Cook, E., Onsted, J., *et al.* (2011). A multi-scalar approach to theorizing socio-ecological dynamics of urban residential landscapes. *Cities and the Environment*, *4*(1), 6.

Clark, T. N., Lloyd, R., Wong, K. K., & Jain, P. (2002). Amenities drive urban growth. *Journal of Urban Affairs*, *24*(5), 493-515.

Couch, C., Petschel-held, G., & Leontidou, L. (2007). *Urban Sprawl in Europe: Landscapes, Land-Use Change and Policy.* Blackwell, London.

Cross, D. (1990). *Counterurbanisation in England and Wales*. Ashgate, Aldershot.

Csefalvay, Z., & Webster, C. (2012). Gates or no gates? A cross-European enquiry into the driving forces behind gated communities. *Regional Studies*, *46*(3), 293-308.

Dalmazzo, A., & De Blasio, G. (2011). Amenities and skill biased agglomeration effects: Some results on Italian cities. *Papers in Regional Science*, *90*(3), 503-527.

Davis, D. E. (2016). Reflections on the relations between development and urbanization: Past trajectories and future challenges. *International Journal of Urban Science*, *20*, 1–14.

Davoudi, S. (2008). Conceptions of the city region: A critical review. *Urban Design and Planning*, *DP2*, 51-60.

Deller, S. C., Tsai, T.-H., Marcouiller, D. W., & English, D. B. K. (2001). The role of amenities and quality of life in rural economic growth. *American Journal of Agricultural Economics*, *83*, 352-365.

Deng, J. S., Wang, K., Hong, Y., & Qi, J. G. (2009). Spatio-temporal dynamics and evolution of land use change and landscape pattern in response to rapid urbanization. *Landscape and Urban Planning*, *92*(3-4), 187-198.

Di Feliciantonio, C., Salvati, L., Sarantakou, E., & Rontos, K. (2018). Class diversification, economic growth and urban sprawl: Evidences from a pre-crisis European city. *Quality & Quantity*, *52*(4), 1501-1522.

Downey, L., & Hawkins, B. (2008). Race, income, and environmental inequality in the United States. *Sociological Perspectives*, *51*(4), 759-781.

Duranton, G., Rodríguez Pose, A., & Sandall, R. (2009). Family types and the persistence of regional disparities in Europe. *Economic Geography*, *85*(1), 23-47.

Durbin, R. A., & Sung, C. (1987). Spatial variation in the price of housing, rent gradients in non-monocentric cities, *Urban Studies*, *24*, 193-204.

Easton, S., Lees, L., Hubbard, P., & Tate, N. (2020). Measuring and mapping displacement: The problem of quantification in the battle against gentrification. *Urban Studies*, *57*(2), 286-306.

Egidi, G., Cividino, S., Vinci, S., Sateriano, A., & Salvia, R. (2020). Towards local forms of sprawl: A brief reflection on Mediterranean urbanization. *Sustainability*, *12*(2), 582.

Elias, P., Dekolo, S., & Babatola, O. (2012). Land use change modelling in developing countries: issues and prospects. *International Journal of Geography and Geology*, *1*(1), 23.

Fielding, A. J. (1982). Counterurbanisation in western Europe. *Progress in Planning*, *17*, 1-52.

Forster, C. (2006). The challenge of change: Australian cities and urban planning in the new millennium. *Geographical Research*, *44*, 173-182.

Fournier, B., Frey, D., & Moretti, M. (2020). The origin of urban communities: From the regional species pool to community assemblages in city. *Journal of Biogeography*, *47*(3), 615-629.

Freeman, L. (2009). Neighbourhood diversity, metropolitan segregation and gentrification: What are the links in the US? *Urban Studies*, *46*(10), 2079-2101.

Frenkel, A., Bendit, E., & Kaplan, S. (2013). Residential location choice of knowledge-workers: The role of amenities, workplace and lifestyle. *Cities*, *35*, 33-41.

Fujita, M., & Ogawa, H. (1982). Multiple equilibria and structural transition of non-monocentric urban configurations. *Regional Science and Urban Economics*, *12*(2), 161-196.

García-Nieto, A. P., Geijzendorffer, I. R., Baró, F., Roche, P. K., Bondeau, A., & Cramer, W. (2018). Impacts of urbanization around Mediterranean cities: Changes in ecosystem service supply. *Ecological Indicators*, *91*, 589-606.

Genis, Ş(2007). Producing elite localities: The rise of gated communities in Istanbul. *Urban Studies*, 44(4), 771-798.

Giannakourou, G. (2005). Transforming spatial planning policy in Mediterranean countries. Europeanization and domestic change. *European Planning Studies*, *13*(2), 253-264.

Gitas, I. Z., Mitri, G. H., & Ventura, G. (2004). Object-based classification for burned area mapping of Creus Cape, Spain, using NOAA-AVHRR imagery. *Remote Sensing of Environment*, *92*(3), 409-413.

Gkartzios, M. (2013). Leaving Athens: Narratives of counterurbanisation in times of crisis. *Journal of Rural Studies*, *32*, 158-167.

González-Val, R., Lanaspa, L., & Sanz-Gracia, F. (2014). New evidence on Gibrat's law for cities. *Urban Studies*, *51*, 93-115.

Gordon, P., & Richardson, H. W. (1996). Beyond polycentricity: The dispersed metropolis, Los Angeles, 1970–1990. *Journal of the American Planning Association*, *62*(3), 289-295.

Gounaridis, D., Chorianopoulos, I., & Koukoulas, S. (2018). Exploring prospective urban growth trends under different economic outlooks and land-use planning scenarios: The case of Athens. *Applied Geography*, *90*, 134-144.

Gourzis, K., Herod, A., Chorianopoulos, I., & Gialis, S. (2021). On the recursive relationship between gentrification and labour market precarisation: Evidence from two neighbourhoods in Athens, Greece. *Urban Studies*, 00420980211031775.

Grossman, G. M., & Krueger, A. B. (1996). The inverted-U: what does it mean? *Environment and Development Economics*, *1*(2), 119-122.

Haase, D., Frantzeskaki, N., & Elmqvist, T. (2014). Ecosystem services in urban landscapes: practical applications and governance implications. *Ambio*, *43*(4), 407-412.

Hadjimichalis, C. (2017). *Crisis Spaces: Structures, Struggles and Solidarity in Southern Europe*. Routledge, England.

Hall, P. (2009). Looking backward, looking forward: the city region of the mid-21st century. *Regional Studies*, *43*(6), 803-817.

Hall, P., & Pain, K. (2006). *The Polycentric Metropolis: Learning from Mega-City Regions in Europe*. Earthscan, London.

Heikinheimo, V., Tenkanen, H., Bergroth, C., Järv, O., Hiippala, T., & Toivonen, T. (2020). Understanding the use of urban green spaces from user-generated geographic information. *Landscape and Urban Planning*, *201*, 103845.

Heynen, N., Perkins, H. A., & Roy, P. (2006). The political ecology of uneven urban green space: The impact of political economy on race and ethnicity in producing environmental inequality in Milwaukee. *Urban Affairs Review*, *42*(1), 3-25.

Humer, A., Cardoso, R., & Meijers, E. (2021). Breaking with the spatial-cycle model: The shift towards 'syncurbanization' in polycentric urban regions. *Regional Studies*, 1-15.

Isik, C., Ongan, S., & Özdemir, D. (2019). The economic growth/development and environmental degradation: Evidence from the US state-level EKC hypothesis. *Environmental Science and Pollution Research*, *26*(30), 30772-30781.

Jia, W., Zhao, S., Zhang, X., Liu, S., Henebry, G. M., & Liu, L. (2021). Urbanization imprint on land surface phenology: The urban–rural gradient analysis for Chinese cities. *Global Change Biology*, *27*(12), 2895-2904.

Jedwab, R., Loungani, P., & Yezer, A. (2021). Comparing cities in developed and developing countries: Population, land area, building height and crowding. *Regional Science and Urban Economics*, *86*, 103609.

Kabisch, N., Strohbach, M., Haase, D., & Kronenberg, J. (2016). Urban green space availability in European cities. *Ecological Indicators*, *70*, 586-596.

Kefalas, G., Kalogirou, S., Poirazidis, K., & Lorilla, R. S. (2019). Landscape transition in Mediterranean islands: The case of Ionian islands, Greece 1985–2015. *Landscape and Urban Planning*, *191*, 103641.

Kim, S. (2020). Inter-municipal relations in city-region governance. *Cities*, *104*, 102771.

Knaus, K., Margies, N., & Schilling, H. (2021). Thinking the city through work. *City*, *25*(3-4), 303-314

Koprowska, K., Łaszkiewicz, E., & Kronenberg, J. (2020). Is urban sprawl linked to green space availability? *Ecological Indicators*, *108*, 105723.

Kostenwein, D. (2021). Between walls and fences: How different types of gated communities shape the streets around them. *Urban Studies*, 0042098020984320.

Istrate, E., Berube, A., & Nadeau, C. A. (2011). Global MetroMonitor https://www.brookings.edu/wp-content/uploads/2016/06/0118_global_metro_monitor.pdf. (accessed 11 December 2021).

Jones, G. W. (2002). Southeast Asian urbanization and the growth of mega-urban regions. *Journal of Population Research*, *19*, 119-136.

Jonas, A. E., & Ward, K. (2007). Introduction to a debate on city-regions: New geographies of governance, democracy and social reproduction. *International Journal of Urban and Regional Research*, *31*(1), 169-178.

Kim, S., & Rowe, P. G. (2012). Does large-sized cities' urbanisation predominantly degrade environmental resources in China? Relationships between urbanisation and resources in the Changjiang Delta Region. *International Journal of Sustainable Development & World Ecology*, *19*(4), 321-329.

Kling, R., Olin, S. C., & Poster, M. (1995). The emergence of post suburbia: An introduction. In R. Kling, S. C. Olin, & M. Poster (Eds.), *Postsuburban California. The Transformation of Orange County Since World War II*, (pp. 1-30). University of California Press, Berkeley.

Kloosterman, R., & Musterd, S. (2001). The polycentric urban region: Towards a research agenda. *Urban Studies*, *38*(4), 623-633.

Krumholz, N. (1992). Roman impressions: Contemporary city planning and housing in Rome. *Landscape and Urban Planning*, *22*(2–4), 107-114.

Kulish, M., Richards, A., & Gillitzer, C. (2012). Urban structure and housing prices: Some evidence from Australian cities. *Economic Record*, *88*(282), 303-322.

La Rosa, D. (2014). Accessibility to greenspaces: GIS based indicators for sustainable planning in a dense urban context. *Ecological Indicators*, *42*, 122-134.

Lawton, P., Murphy, E., & Redmond, D. (2013). Residential preferences of the 'creative class'? *Cities*, *31*, 47-56.

Leal Filho, W., Barbir, J., Sima, M., Kalbus, A., Nagy, G. J., Paletta, A.,... & Bonoli, A. (2020). Reviewing the role of ecosystems services in the sustainability of the urban environment: A multi-country analysis. *Journal of Cleaner Production*, *262*, 121338.

Le Goix, R. (2005). Gated communities: Sprawl and social segregation in Southern California. *Housing Studies*, *20*(2), 323-343.

Lever, W. F. (1993). Reurbanisation—The policy implications. *Urban studies*, *30*(2), 267-284.

Liu, Y., Yang, L., & Jiang, W. (2020). Coupling coordination and spatiotemporal dynamic evolution between social economy and water environmental quality–A case study from Nansi Lake catchment, China. *Ecological Indicators*, *119*, 106870.

Liu, Y., Zeng, C., Cui, H., & Song, Y. (2018). Sustainable land urbanization and ecological carrying capacity: A spatially explicit perspective. *Sustainability*, *10*(9), 3070.

Marquard, E., Bartke, S., Gifreu I. Font, J., Humer, A., Jonkman, A., Jürgenson, E., Marot, N., Poelmans, L., Repe, B., Rybski, R., Schröter-Schlaack, C., Sobocká, J., Sørensen, M. T., Vejchodská, E., Yiannakou, A., & Bovet, J. (2020). Land consumption and land take: Enhancing conceptual clarity for evaluating Spatial Governance in the EU context. *Sustainability*, *12*(19), 8269.

Martinez-Fernandez, C., Audirac, I., Fol, S., & Cunningham-Sabot, E. (2012). Shrinking cities: Urban challenges of globalization. *International Journal of Urban and Regional Research*, *36*, 213-225.

McCann, P. (2017). Urban futures, population ageing and demographic decline. *Cambridge Journal of Regions, Economy and Society*, *10*, 543-557.

McCann, E. J. (2007). Inequality and politics in the creative city-region: Questions of liveability and state-strategy. *International Journal of Urban and Regional Research*, *31*(1), 188-196.

MEA (Millennium Ecosystem Assessment). (2005a). *Ecosystems and Human Well-Being: Synthesis Report World Resources Institute*. Island Press, Washington D.C., p. 137.

MEA (Millennium Ecosystem Assessment) (Program). (2005b). *Ecosystems and Human Well-Being: Synthesis*. Island Press, Washington, D.C.

Meijers, E., & Sandberg, K. (2021). Territorial cohesion, polycentrism and regional disparities: Revisiting an unsolved debate. In D. Rauhut, S. Sielker, & A. Humer (Eds.), *EU Cohesion Policy and Spatial Governance: Territorial, Social and Economic Challenges*, (pp. 127–141). Edward Elgar, Cheltenham.

Meijers, E., Hoogerbrugge, M., & Cardoso, R. (2018). Beyond polycentricity: Does stronger integration between cities in polycentric urban regions improve performance? *Tijdschrift voor Economische en Sociale Geografie*, *109*(1), 1-21.

Menke, T. (1987). Economic welfare and urban amenities across race-sex groups. *Urban Studies*, *24*(2), 151-161.

Mingione, E. (2009). Family, welfare and districts: The local impact of new migrants in Italy. *European Urban and Regional Studies*, *16*(3), 225-236.

Mohai, P., & Saha, R. (2006). Reassessing racial and socioeconomic disparities in environmental justice research. *Demography*, *43*(2), 383-399.

Mordridge, M., & Parr, J. (1997). Metropolis or region: On the development and structure of London. *Regional Studies*, *31*, 97-115.

Munafò, M., Norero, C., Sabbi, A., & Salvati, L. (2010). Urban soil consumption in the growing city: A survey in Rome. *Scottish Geographical Journal*, *126*(3), 153-161.

Musterd, S., Marcińzak, S., Van Ham, M., & Tammaru, T. (2017). Socioeconomic segregation in European capital cities. Increasing separation between poor and rich. *Urban Geography*, *38*(7), 1062-1083.

Nor, A. N. M., Corstanje, R., Harris, J. A., & Brewer, T. (2017). Impact of rapid urban expansion on green space structure. *Ecological Indicators*, 81, 274-284.

Nyström, J. (1992). The cyclical urbanization model: A critical analysis. *Geografiska Annaler: Series B, Human Geography*, *74*(2), 133-144.

Parker, J., & Simpson, G. D. (2018). Public green infrastructure contributes to city livability: A systematic quantitative review. *Land*, 7(4), 161.

Parr, J. (2012). The spatial-cycle model (SCM) revisited. *Regional Studies, 46*(2), 217-228. https://doi.org/10.1080/00343404.2011.558895

Paulsen, K. (2012). Yet even more evidence on the spatial size of cities: Urban spatial expansion in the US, 1980–2000. *Regional Science and Urban Economics, 42*(4), 561-568.

Petanidou, T., Kizos, T., & Soulakellis, N. (2008). Socioeconomic dimensions of changes in the agricultural landscape of the Mediterranean basin: A case study of the abandonment of cultivation terraces on Nisyros Island, Greece. *Environmental Management, 41*(2), 250-266.

Pham, T., Apparicio, P., Séguin, A. M., Landry, S., & Gagnon, M. (2012). Spatial distribution of vegetation in Montreal: An uneven distribution or environmental inequity? *Landscape and Urban Planning, 107*(3), 214-224.

Phelps, N. A., & Wu, F. (Eds.). (2011). *International Perspectives on Suburbanization*. Palgrave Macmillan, London.

Rappaport, J. (2008). Consumption amenities and city population density. *Regional Science and Urban Economics, 38*(6), 533-552.

Rauch, J. E. (1991). *Economic Development, Urban Underemployment, and Income Inequality*.

Rauhut, D., & Humer, A. (2020). EU cohesion policy and spatial economic growth: Trajectories in economic thought. *European Planning Studies, 28*(11), 2116-2133. https://doi.org/10.1080/09654313.2019.1709416

Reher, D. S. (2011). Economic and social implications of the demographic transition. *Population and Development Review, 37*, (Suppl. 1), 11–33.

Reisig, D., Mullan, K., Hansen, A., Powell, S., Theobald, D., & Ulrich, R. (2021). Natural amenities and low-density residential development: Magnitude and spatial scale of influences. *Land Use Policy, 102*, 105285.

Rickards, L., Gleeson, B., Boyle, M., & O'Callaghan, C. (2016). Urban studies after the age of the city. *Urban Studies, 53*(8), 1523-1541.

Rodriguez-Pose, A. (2008). The rise of the 'city-region' concept and its development policy implications. *European Planning Studies, 16*(8), 1025-1046.

Rogers, A. (1982). Sources of urban population growth and urbanization, 1950–2000: A demographic accounting. *Economic Development and Cultural Change, 30*, 483-506.

Saganeiti, L., Favale, A., Pilogallo, A., Scorza, F., & Murgante, B. (2018). Assessing urban fragmentation at regional scale using sprinkling indexes. *Sustainability*, *10*(9), 3274.

Salvati, L. (2015). Agro-forest landscape and the 'Fringe' City: A multivariate assessment of land-use changes in a sprawling region and implications for planning. *Science of the Total Environment 490*, 715-723.

Salvati, L. (2014). Towards a polycentric region? The socioeconomic trajectory of Rome, an 'ethernal' Mediterranean city. *Tijdschrift voor Economische en Sociale Geographie*, *105*(3), 268-284.

Salvati, L., & Gargiulo Morelli, V. (2014). Unveiling urban sprawl in the Mediterranean region: Towards a latent urban transformation? *International Journal of Urban and Regional Research*, *38*(6), 1935-1953.

Salvati, L., Zitti, M., & Sateriano, A. (2013). Changes in the city vertical profile as an indicator of sprawl: Evidence from a Mediterranean region. *Habitat International*, *38*, 119-125.

Sayadi, S., González-Roa, M. C., & Calatrava-Requena, J. (2009). Public preferences for landscape features: The case of agricultural landscape in mountainous Mediterranean areas. *Land Use Policy*, 26(2), 334-344.

Schwarz, K., Berland, A., & Herrmann, D. L. (2018). Green, but not just? Rethinking environmental justice indicators in shrinking cities. *Sustainable Cities and Society*, *41*, 816-821.

Scott, A. J. (2019). City-regions reconsidered. Environment and Planning A. *Economy and Space*, *51*(3), 554-580. https://doi.org/10.1177/0308518X19831591

Segal, D. (1976). Are there returns to scale in city size? *The Review of Economics and Statistics*, 339-350.

Seto, K. C., Frankians, M., Güneralp, B., & Reilly, M. K. (2011). A meta-analysis of global urban land expansion. *PLoS ONE*, *6*, e23777.

Shrestha, M. K., York, A. M, Boone, C. G., & Zhang, S. (2012). Land fragmentation due to rapid urbanization in the Phoenix Metropolitan area: Analyzing the spatiotemporal patterns and drivers. *Applied Geography*, *32*, 522-531.

Shu, X., Han, H., Huang, C., & Li, L. (2019). Defining functional polycentricity from a Geographical perspective. *Geographical Analysis*. https://doi.org/10.1111/gean.12195

Silva, P. (2020). Not so much about informality: Emergent challenges for urban planning and design education. *Sustainability*, *12*(20), 8450.

Spivey, C. (2008). The Mills—Muth model of urban spatial structure: Surviving the test of time? *Urban Studies*, *45*(2), 295-312.

Stephens, H. M., & Partridge, M. D. (2015). Lake amenities, environmental degradation, and great lakes regional growth. *International Regional Science Review*, *38*(1), 61-91.

Ströbele, M., & Hunziker, M. (2017). Are suburbs perceived as rural villages? Landscape-related residential preferences in Switzerland. *Landscape and Urban Planning*, *163*, 67-79.

Takahashi, T. (2014). Location competition in an Alonso–Mills–Muth city. *Regional Science and Urban Economics*, *48*, 82-93.

Trounstine, J. (2020). The geography of inequality: How land use regulation produces segregation. *American Political Science Review*, *114*(2), 443-455.

Turok, I., & Mykhnenko, V. (2007). The trajectories of European cities, 1960-2005. *Cities*, *24*, 165-182.

Van Criekingen, M. (2010). 'Gentrifying the re urbanisation debate', not vice versa: The uneven socio spatial implications of changing transitions to adulthood in Brussels. *Population, Space and Place*, *16*(5), 381-394.

Van de Kaa, D. J. (1987). Europe's second demographic transition. *Population Bulletin*, *42*(1), 1-59.

Van den Berg, L., Drewett, R., Klaassen, L. H., Rossi, A., & Vijverberg, C. H. T. (1982). *Urban Europe: A Study of Growth and Decline*. Pergamon Press, Oxford.

Van Meeteren, M. (2016). From Polycentricity to Renovated Urban Systems Theory: Explaining Belgian Settlement Geographies (Doctoral dissertation). Ghent University, Ghent.

Venter, Z. S., Shackleton, C. M., Van Staden, F., Selomane, O., & Masterson, V. A. (2020). Green apartheid: Urban green infrastructure remains unequally distributed across income and race geographies in South Africa. *Landscape and Urban Planning*, *203*, 103889.

Vergou, P. (2019). Living with difference: Refugee education and school segregation processes in Greece. *Urban Studies*, *56*(15), 3162-3177.

WBGU. (2016). *Humanity on the Move: Unlocking the Transformative Power of Cities*. WBGU, Berlin.

Weber, H., & Sciubba, J. D. (2018). The effect of population growth on the environment: Evidence from European regions. *European Journal of Population*, *35*, 379-402.

White, M. J. (1999). Urban areas with decentralized employment: Theory and empirical work. *Handbook of Regional and Urban Economics*, *3*, 1375-1412.

Wilson, C. (2001). On the scale of global demographic convergence 1950–2000. *Population Development Review*, *27*, 155-171.

Wu, J. (2006). Environmental amenities, urban sprawl, and community characteristics. *Journal of Environmental Economics and Management*, *52*(2), 527-547.

Wüstemann, H., Kalisch, D., & Kolbe, J. (2017). Access to urban green space and environmental inequalities in Germany. *Landscape and Urban Planning*, *164*, 124-131.

Yang, Y., Liu, Y., Li, Y., & Du, G. (2018). Quantifying spatio-temporal patterns of urban expansion in Beijing during 1985–2013 with rural-urban development transformation. *Land Use Policy*, *74*, 220-230.

Ye, V. Y., & Becker, C. M. (2018). The (literally) steepest slope: spatial, temporal, and elevation variance gradients in urban spatial modelling. *Journal of Economic Geography*, *18*(2), 421-460.

Yuan, L., Shin, K., & Managi, S. (2018). Subjective well-being and environmental quality: The impact of air pollution and green coverage in China. *Ecological Economics*, *153*, 124-138.

Zhou, X., & Wang, Y. C. (2011). Spatial–temporal dynamics of urban green space in response to rapid urbanization and greening policies. *Landscape and Urban Planning*, *100*(3), 268-277.

Zimmermann, K., & Atkinson, R. (2021). Urban policy in European Cohesion policy. In D. Rauhut, S. Sielker, & A. Humer (Eds.), *EU Cohesion Policy and Spatial Governance: Territorial, Social and Economic Challenges*, (pp. 83–97). Edward Elgar, Cheltenham.

Zitti, M., Ferrara, C., Perini, L., Carlucci, M., & Salvati, L. (2015). Long-term urban growth and land-use efficiency in Southern Europe: Implications for sustainable land management. *Sustainability*, *7*, 3359-3385.

# 3

# Land Resource Consumption in a Tourism Destination: Long-term Dynamics in Mykonos, Greece

**Gianluigi Salvucci[1], Marco Zitti[2], and Kostas Rontos[3]**

[1]Italian National Institute of Statistics (ISTAT), Italy
[2] c/o Italian Council of Agricultural Research and Economics, Italy
[3]Department of Sociology, University of the Aegean, Greece

## Abstract

Peri-urban sprawl in Southern Europe is a consolidated process throughout Europe causing (more or less important) processes of land resource consumption. The intrinsic characteristics of regions undergoing sprawl are variable along the Mediterranean basin and no common features can be identified based on the present information systems, especially from official statistics. The peculiar attributes of Mediterranean sprawl are difficult to assess due to the restricted availability of comparable statistical and detailed maps. The comparative analysis of case studies that differ in population size and density, socioeconomic features, land availability, and region may provide sound information to overcome the mentioned information gap. Mykonos, Greece is an example of tourism-driven sprawl occurring in an originally low-density area far outside the direct influence of urban regions. The assessment of Mykonos sprawl was carried out by integrating statistical data from multiple sources, including the censuses of population, buildings, and agriculture, land-use and cadastral maps, as well as a photographic field survey. Results (i) define the different phases of urban sprawl in Mykonos, (ii) illustrate the peculiarity of this process compared to other areas undergoing sprawl in Greece, and, finally, (iii) represent an assessment framework for monitoring a so complex and differentiated phenomenon.

## 3.1 Introduction

Urban sprawl in Southern Europe was affected by various factors including elevation, population density, settlement patterns, closeness to the sea, pristine land use, and land availability. In accessible, flat areas along sea coasts, sprawl has intensified in recent years (Leontidou and Marmaras, 2001) following to the consolidation of city networks with service-oriented economies and tourism development (Alphan, 2003; Polyzos *et al.*, 2008; Kucukmehmetoglu and Geymen, 2009). The negative impact of sprawl in terms of soil consumption and landscape fragmentation needs further documentation in Mediterranean Europe (Bruegmann, 2005). Interestingly enough, this process was recently spreading across rural areas progressively further away from urban regions. Tourism resorts, delocalized commercial and industrial infrastructures, and dense road networks, coupled with the conversion of isolated buildings from agricultural to residential uses are the most visible marks of the transition from rural to peri-urban territories (Couch *et al.*, 2007).

Permanent monitoring of sprawl and the consequent land consumption outside large urban regions is, therefore, a crucial task in sustainable regional development (European Environmental Agency, 2006). Several indicators were introduced with the aim of measuring urban sprawl in (originally) rural regions (e.g., Hasse and Lathrop, 2003; Frenkel and Ashkenazi, 2007; Schneider and Woodcock, 2008). Some of them were implemented with the aim of evaluating diachronically the effectiveness of management strategies in soil and land protection (Bruegmann, 2005). Unfortunately, the computation of such indicators is time-consuming and sometimes needs important technical resources (Couch *et al.*, 2007).

The case study illustrated in this chapter (Mykonos island in the archipelago of Kyklades, Greece) is a paradigmatic example of the process of sprawl occurring spatially far away from (and functionally outside the direct influence of) urban regions. The assessment of Mykonos land consumption due to sprawl was carried out by integrating statistical data from multiple sources: the censuses of population, buildings, and agriculture, land-use and cadastral maps, as well as a field survey. Results illustrate the different phases of Mykonian sprawl and indicate some potential causes of this complex process.

## 3.2 Methodology

### 3.2.1 Study area

The research area covers the whole island of Mykonos (Kyklades, Greece) with a surface area of 85 km$^2$ (Figure 3.1). As one of the most traditional tourism centres in the Mediterranean, the island underwent important land-use changes in the last 50 years due to compact growth of the main centre (Mykonos town). Tourism and second-home development in suburban areas have also caused concentration of people and buildings especially along the coastal rim (Coccossis *et al.*, 1996). Using the results obtained from family reconstitution and inverse projection, Hionidou (1995) examined the demographic dynamics in Mykonos (1859–1959). The pre-transitional demographic system of Mykonos was found to have been characterized by the combination of low mortality and low marital though moderately high overall fertility. The resulting high population growth was counterbalanced by high emigration which kept the island population almost constant. Mortality and fertility experienced parallel and rapid decline. In both pre-transitional and transitional Mykonos, fertility rather than mortality seems to have been the main factor responsible for the changes in population growth. The findings point to a demographic distinctiveness of Mediterranean island populations.

### 3.2.2 Statistical analysis

Land-use distribution was assessed in Mykonos and the Kyklades from 1960 to 2000. Data were obtained from the land-use census (available years: 1960, 1970, 1980, and 1990) carried out by the Greek National Statistical Office (ELSTAT) together with the National Census of Agriculture and Livestock and the CORINE land cover project (available years: 1990 and 2000). Six basic categories of land uses were investigated in the land-use census: cultivated surfaces, pastures, forests, inland water, urban areas, and other unproductive

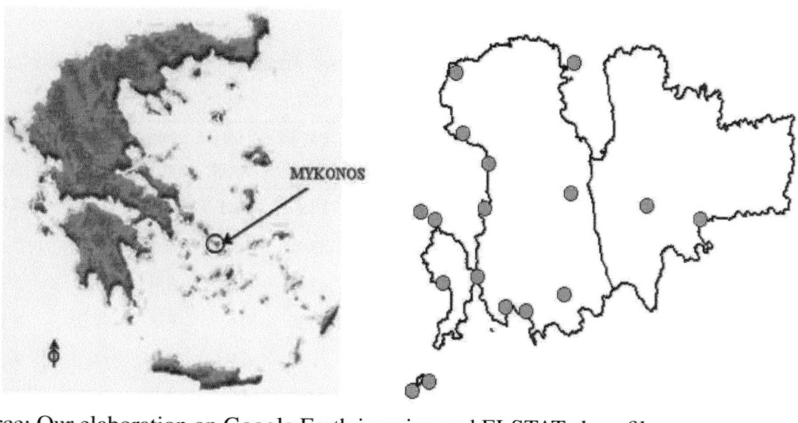


Source: Our elaboration on Google Earth imagine and ELSTAT shapefiles.

**Figure 3.1** The map of Greece with the location of Mykonos (left) and the location of villages in Mykonos island (right).

lands (Salvati et al., 2009). Land-use composition was calculated in each spatial domain as the per cent surface area of each land-use class on the total investigated area. Further information on settlement density was obtained from the 2000 census of buildings (number and surface of buildings by year of construction and municipality or building block). Population census was used to derive resident and usual population occurring in each enumeration district during the investigated period. All data were provided by the ELSTAT.

## 3.3 Results

A complex demographic pattern was observed in the investigated area. In the most recent period, population grew from 6000 resident people in 1991 to more than 9000 people in 2001. Although the growing rate of the resident population amounted to 5% *per year*, population density remains relatively low (about 110 inhabitants km$^{-2}$). Until 1970s, the island of Mykonos was characterized by a traditional spatial organization with compact settlements concentrated within the area of the main city and the harbour. Such a settlement model is similar to that observed in several other Greek islands and reflects the traditional compactness of Mediterranean urban areas and coastal towns. In the following years, an impressive growth of dispersed settlements (especially one-floor houses with small gardens or two-floor residential housing estates) was observed (Table 3.1). Density of buildings continuously grew between 1919 and 2000 and amounts to nearly 1.2 buildings *per hectare* in the most recent period. This density is larger than that observed in the whole Kyklades prefecture (0.4 buildings *per hectare*).

**Table 3.1** Settlement date of construction in various areas of Greece (per cent share of buildings).

| Region | < 1919 | 1919–1945 | 1945–1960 | 1961–1970 | 1971–1980 | 1981–1985 | 1986–1990 | 1991–1995 | 1996–2000 |
|---|---|---|---|---|---|---|---|---|---|
| **Greece** | **5.0** | **10.2** | **16.7** | **19.1** | **18.5** | **10.1** | **7.5** | **6.1** | **4.8** |
| *Attica* | *1.0* | *4.5* | *14.3* | *21.4* | *23.8* | *11.7* | *8.5* | *6.7* | *5.5* |
| Kyklades | 15.7 | 12.7 | 12.2 | 11.1 | 12.7 | 9.6 | 8.9 | 8.0 | 5.2 |
| Mykonos (island) | 11.9 | 11.4 | 6.0 | 13.0 | 19.2 | 8.9 | 9.7 | 10.2 | 8.8 |
| *Mykonos town* | *16.3* | *15.5* | *7.5* | *14.7* | *19.4* | *6.9* | *7.2* | *6.9* | *4.7* |
| *The remaining settlements* | *0.2* | *3.5* | *2.5* | *8.1* | *14.7* | *19.7* | *21.9* | *17.3* | *11.7* |

Source: Elaboration on ELSTAT, Census of Buildings.

**Table 3.2** A comparison of agricultural systems by land-use class in Kyklades archipelago (Aegean Sea), Attica region (Central Greece) and the country as a whole.

| Land cover categories | Greece | Attica | Kyklades | Thira | Kythnos | Milos | Mykonos | Naxos | Paros | Sifnos | Kea | Serifos | Sikinos |
|---|---|---|---|---|---|---|---|---|---|---|---|---|---|
| Arable land | **16.0** | *1.8* | 10.8 | 0.0 | 15.6 | 16.0 | *28.9* | 0.9 | 32.1 | 10.0 | 19.6 | 13.8 | 0.0 |
| Permanent crops | **5.7** | *6.3* | 1.8 | 0.0 | 0.0 | 0.8 | *0.0* | 2.0 | 1.3 | 5.2 | 14.9 | 0.0 | 0.0 |
| Pastures-transitional wood land/shrub | **0.7** | *0.2* | 0.2 | 0.0 | 0.0 | 0.0 | *0.0* | 0.0 | 0.0 | 0.0 | 0.0 | 0.0 | 0.0 |
| Pastures-shrub/herbaceous vegetation | **6.9** | *4.8* | 26.2 | 11.4 | 67.1 | 26.1 | *26.5* | 17.1 | 23.3 | 15.5 | 18.8 | 54.4 | 0.0 |
| Pastures-Open spaces, little vegetation | **3.3** | *0.8* | 4.0 | 0.0 | 3.8 | 0.1 | *11.8* | 0.0 | 0.6 | 2.3 | 17.3 | 3.4 | 0.0 |
| Heterogeneous agricultural areas | **16.7** | *23.3* | 19.6 | 53.3 | 2.8 | 17.2 | *9.5* | 50.1 | 19.6 | 17.2 | 20.6 | 3.5 | 0.0 |

Source: Elaboration on ELSTAT, Land-use census.

Long-term land-use changes in Mykonos involved primarily agricultural lands and pastures. In 1960, the agricultural land surface was 18% of the whole research area and fell to 15% in 1990. Pastures showed the reverse pattern. At that time, land use in Mykonos was peculiar compared with the other Kyklades islands (Table 3.2). Urban areas and other "unproductive lands" amounted to 46% of total island surface (compared to 6% observed, on average, in Kyklades) and woodlands were completely absent (2% in Kyklades). A similar pattern was observed in 2000. Based on the analysis of Corine land cover maps, it was found that the artificial surfaces are more common in Mykonos (2%) than in the whole Kyklades prefecture (1%). The percentage observed in Mykonos is similar to (or even higher than) that observed in other large islands with high population density and growth and important tourism pressure (e.g., Naxos, Paros, Syros, and Thira). Apart from arable lands, the open spaces with little or no vegetation are the predominant land cover in Mykonos (more than 23% of the total island surface compared to 13% observed in Kyklades) overall indicating a low-diversity landscape. Seasonal population growth and urban sprawl, thus, impact a very fragile and poor landscape and may trigger soil and landscape degradation.

## 3.4 Discussion

This contribution offers an example of integrated monitoring to apply in rural, low-density areas undergoing land-use changes. The recent development phase in Mykonos was associated with important processes of land consumption. The conversion of agricultural and pasture land located in the suburbs of the main centre, fuelled by important socio-economic changes, was driven by tourism and house speculation. After World War II, the lack of urban planning as well as poor control against land appropriation and illegal buildings were the main causes of land consumption, with a similar pattern with other Mediterranean islands that act as tourism poles (Leontidou and Marmaras, 2001). On the contrary, these dynamics are different in respect to the other Kyklades islands because of the different settlement patterns and landscape composition.

In Mykonos, urban sprawl concentrated (i) on the western zone of the island, (ii) close to the main town and the airport, and (iii) along the main beaches. Notably, these newly developed areas can be classified as a mix of service (e.g., commercial, industrial, and transport) infrastructures and residential (second-home) settlements with the open spaces originally devoted to agriculture and then converted to gardens or even abandoned. From the environmental point of view, they represent "interstitial" spaces with poor ecological quality, possibly waiting for further conversion to urban uses. The future of the open spaces close to the diffused, low-density settlements could be land degradation and habitat fragmentation due to urbanization. These spaces will be "spontaneously" colonized by housing estates as well as services (such as artisans, car sellers, and small factories) replacing the old farm-holders and potentially increasing the "sprawl" process and soil sealing rate (Karatzoglou and Spilanis, 2010). Trends in *per-capita* sealed surface have been evaluated by comparing over time the trend in soil consumption with data on resident population (e.g., Mazzetti, 2007). According to this indicator, a "decoupling" process between sprawl and population growth was observed during the investigated period. While soil consumption increased at a very high rate, population density was relatively stable confirming the importance of the recent (second-home-driven and tourism-driven) urbanization.

## 3.5 Concluding Remarks

The issue of urban sprawl leading to soil sealing in island regions is becoming crucial in the European research and policy agenda. In Southern Europe, several research institutions, environmental societies, and local groups started studying and divulgating the negative impact of these processes on the intrinsically fragile landscape of the Mediterranean islands. This problem claims for renewed monitoring efforts based on the integration of statistical surveys, remote sensing, and field mapping (e.g., Weber *et al.*, 2005). The present study is a preliminary contribution to this issue and confirms the crucial role of official statistics databases (e.g., census of buildings, land-use census, population census, building activity statistics, agricultural census, and annual agricultural statistics) to assure a timely and spatially complete assessment of sprawl within and especially outside urban regions.

## References

Alphan, H., Land use change and urbanisation of Adana, Turkey, *Land Degradation and Development*, 14, 575-586, (2003).

Bruegmann, R., *Sprawl: A Compact History*, Chicago, University of Chicago Press, (2005).

Calafati, A., Urban sprawl, Italian style, *Italian Journal of Regional Science*, 5, 28-36, (2008).

Coccossis, H., Parpairis, A. & Priestley, G. K., Tourism and carrying capacity in coastal areas: Mykonos, Greece, in: Priestley, G. K., Edwards, J. A., and Coccossis, H. (Eds.), *Sustainable Tourism? European Experiences*, pp. 153-175, (1996).

Couch, C., Petschel-held, G. & Leontidou, L., *Urban Sprawl in Europe: Landscapes, Land-Use Change and Policy*, London, Blackwell, (2007).

European Environmental Agency, *Urban Sprawl in Europe – The Ignored Challenge*, Report no. 10, Copenhagen, (2006).

Frenkel, A. & Ashkenazi, M., The integrated sprawl index: measuring the urban landscape in Israel, *Annals of Regional Science*, 42(1), 99-121, (2007).

Hasse, J. E. & Lathrop, R. G., Land resource impact indicators of urban sprawl, *Applied Geography*, 23, 159-175, (2003).

Hionidou, V., The demographic system of a Mediterranean Island: Mykonos, Greece 1859–1959, *International Journal of Population Geography*, 1(2), 125–146, (1995).

Karatzoglou, B. & Spilanis, I., Sustainable tourism in Greek islands: the integration of activity-based environmental management with destination environmental scorecard based on the adaptive resource management paradigm, *Business Strategy and the Environment*, 19(1), 26-38, (2010).

Kucukmehmetoglu, M. & Geymen, A., Urban sprawl factors in the surface water resource basins of Istanbul. *Land Use Policy*, 26, 569-579, (2009).

Leontidou, L., *The Mediterranean City in Transition*, Cambridge, Cambridge University Press, (1990).

Leontidou, L. & Marmaras, E., From tourists to migrants: residential tourism and 'littoralisation', in: Apostolopoulos., Y., Loukissas, P., and Leontidou, L. (Eds.), *Mediterranean Tourism. Facets of Socioeconomic Development and Cultural Change*, New York, Routledge, (2001).

Mazzetti, E., Radiografia di un organismo geografico complesso: l'isola di Capri, *Rivista Geografica Italiana*, 114(4), 593-619, (2007).

Polyzos, S., Christopoulou, O., Minetos, D. & Leal Filho, W., An overview of urban-rural land use interactions in Greece, *International Journal of Agricultural Resources, Governance and Ecology*, 7, 276-296, (2008).

Salvati, L., Sabbi, A., Gargiulo, V., Rontos, K. & Benaki V., Exploring land use relationships in a Mediterranean city region, *Statistical Review*, 5(1-2), 103-124, (2009).

Schneider, A. & Woodcock, C. E., Compact, dispersed, fragmented, extensive? A comparison of urban growth in twenty-five global cities using remotely sensed data, pattern metrics and census information, *Urban Studies*, 45(3), 659-692, (2008).

Weber, C., Petropoulou, C. & Hirsch, J., Urban development in the Athens metropolitan area using remote sensing data with supervised analysis and GIS. *International Journal of Remote Sensing*, 26(4), 785-796, (2005).

# 4

# Beyond Crisis: Demographic Dynamics, Long-term Settlement Patterns, and the Intrinsic Evolution of Local Systems Along a Coastal and Inland Gradient

**Kostas Rontos[1], Vito Imbrenda[2], Rosanna Salvia[3], and Giovanni Quaranta[3]**

[1]Department of Sociology, University of the Aegean, Greece
[2]Institute of Methodologies for Environmental Analysis, National Research Council of Italy (IMAA-CNR), 85050, Tito Scalo, Italy
[3]Department of Mathematics, Computer Science and Economics, University of Basilicata, Potenza, Italy

## Abstract

This work focuses on population trends recorded at municipal scale in Latium, Central Italy, in the time period between 1871 and 2011. A principal component analysis (PCA) and a cluster analysis were adopted to investigate demographic dynamics concerning specifically the development and the subsequent consolidation of a coastal-inland gradient in population density within the examined area. The main findings of the study indicate the prominence of internal migrations occurred both along elevation (from mountainous to flat areas) and urban (from Rome to countryside) gradients. Population displacement has generated a socioeconomic gap between lowlands featuring intensive agriculture and inner areas featuring depopulation and land abandonment. Peri-urbanization and coastalization of economic activities determined remarkable changes in population density,

[1] University of the Aegean, Department of Sociology, University Hill, GR-8100 Mitilini, E-mail: k.rostos@soc.aegean.gr

amplifying traditional divides in the region. Our study demonstrates the role of a multivariate, exploratory approach in the study of regional population dynamics.

## 4.1 Introduction

The drivers of changes in population dynamics and landscape arrangement are certainly attributable to variable economic processes occurring both at global and local level (Herrmann and Hutchinson, 2005). The unprecedented growth of population registered in the last 60 years, reaching alarming levels since the 1990s, has stimulated scientists and decision makers to activate to reconcile economic development and environmental challenges (Glenn *et al.*, 1998). In fact, the overall mismatch observed between the economic carrying capacity and the demographic density of a given (developed or developing) country is attributable to population movements.

The locution "carrying capacity" offers a dual interpretation: from an economic point of view, it is the capability of an economy to sustain a given population (Knerr, 1998; Goria, 2000), and from an ecological point of view, it can be considered as the maximum number of people able to live in a given area preserving a healthy lifestyle in the long-term by exploiting the pool of the available resources (Blaikie and Brookfield, 1987). In practical terms, carrying capacity is linked to the natural capital available, the human resources living in the investigated area, the technological level of the region, the governance and the institutional set up, the complex heritage of cultural and social aspects, and the current possibility to exchange goods and services within the global community (Goria, 2000).

Generally speaking, positive demographic trends observed in a district for many years imply the need of an increasing carrying capacity (Cuffaro, 2001). If this does not happen, the only social solution is emigration. Population displacements are seen as instability drivers, representing a social threat also for the repercussions on the quality of environments (Chopra and Gulati, 1997). This dynamics is particularly amplified when areas under investigation are coastal zones. Here, the socio-demographic changes take form of heavy and chaotic urbanization processes (sprawl), internal migration, and littoralization (concentration of economic activities along coastal areas), often accompanied by an overexploitation of natural resources (Knerr, 1998;

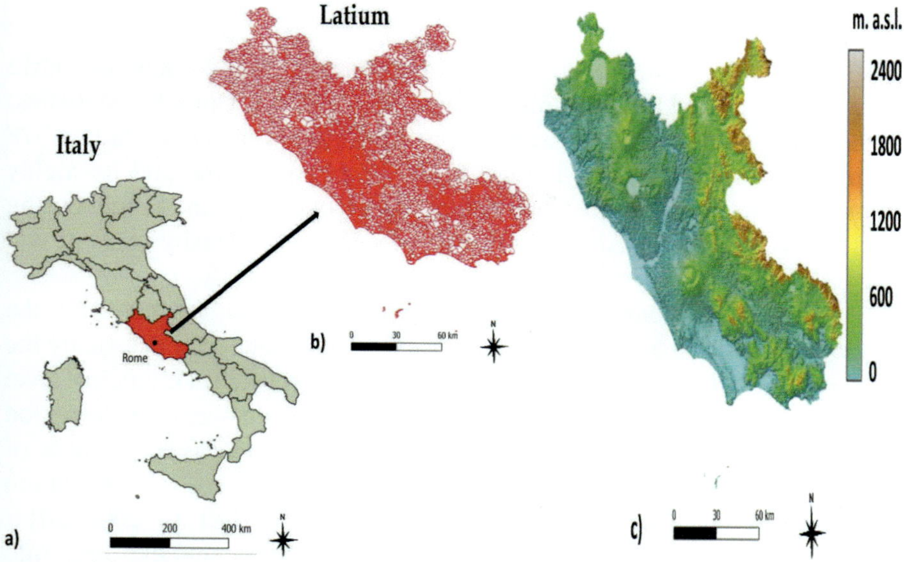


**Figure 4.1** (a) Italy divided into 20 regions (NUTS2); (b) Latium divided into 378 municipalities; (c) digital elevation model (DEM) with a spatial resolution of 20 m.

Garcia Latorre *et al.*, 2001; Tanrivermis, 2003; Imbrenda *et al.*, 2018). The other side of the coin is the process of depopulation, associated with the inexorable ageing of farmers, characterizing traditional (extensive) crop systems, and generating land abandonment and subsequent hydrogeological risks (mainly soil erosion) in rugged environments (Garcia-Ruiz *et al.*, 2011; Tarolli *et al.*, 2014; Agnoletti *et al.*, 2020; Boccia *et al.*, 2020; Quaranta *et al.*, 2020).

Based on these premises, the present study aims at understanding population dynamics of Latium (Central Italy) at municipal level in a wide time window (1871–2011) by using a multivariate exploratory technique. Specifically, the analysis wants to evidence the central role assumed by internal migration flows, occurred along both elevation (from mountains to lowlands) and urban–rural gradients (from the main urban centres to the surrounding countryside). The working hypothesis formulated in our study is that population displacement occurred in Latium in 140 years has impacted mostly coastal zones and lowland areas with intensive agriculture more than mountainous and inner areas widening the existing socioeconomic disparities.

## 4.2 Methodology

The area under investigation is Latium, one of the 20 administrative regions of Italy (NUTS2 – Nomenclature of Territorial Units for Statistics) located in the heart of Italy, extending nearly 17.000 $km^2$ and including 378 municipalities (Figure 4.1). Latium hosts the city of Rome and its highly populated metropolitan area that are in stark contrast to the neighbouring rural areas, characterized by low levels of population density (Salvati and Zitti, 2005).

The primary sources of information for population estimates are the General Population and Household Censuses carried out each decade by the National Institute of Statistics (ISTAT). Data are aggregated at municipal level to provide population density and demographic variation over the time period analysed (1871–2011). Population density was computed as the number of inhabitants per square kilometres. We also mapped the spatial distribution of population in each analysed decade from 1871–1881 to 2001–2011. According to the guidelines of ISTAT (1958), we segmented elevation into four classes: lowlands, coastal hilly areas, internal hilly areas, and mountain. These classes were used to analyse in depth population data. The peculiar feature of lowland municipalities is that they are located at altitudes between 0 and 100 m above sea level (a.s.l.) with a maximum elevation lower than 300 m. Coastal hilly areas have a mean elevation ranging between 100 and 600 m a.s.l. and are located in the proximity of the coastline (Tyrrhenian Sea). Internal hilly areas belong to the same altitudinal belt of the previous class, but they differ in their geographical position (distance from the sea higher than 10 km). Lastly, mountain municipalities are characterized by a mean elevation more than 600 m at the sea level.

We applied a principal components analysis (PCA) on population density dataset to capture an in-depth outlook of population dynamics at municipal scale. The number of significant axes was established according to the scree-plot. In addition to the PCA, we performed a non-hierarchical cluster analysis (i.e., *k*-means) on population density changes occurred in the investigated area in the analysed period. Findings were analysed in the light of three variables: elevation, distance to sea, and a dummy parameter corresponding to the dualism urban–rural. All these elaborations are aimed at generating a spatial assessment of the anthropic pressure on the investigated municipalities according to the distribution of population along the elevation gradient.

## 4.3 Results

As shown in Table 4.1, demographic dynamics of Latium in the period 1871–2011 are consistent. As a first fact, the mean population density increased from 80 inhabitants/km$^2$ in 1871 to more than 170 inhabitants/km$^2$ in 2011, with an accelerate rate in the time window 1881–1951 and especially in the 20 years period 1881–1901 (+21%). It is worth noting that the sharp increase observed in lowlands in the time frame 1931–1936 was caused by land reclamation works in the Fascist period. These were concentrated in the areas of Agro Pontino and Agro Romano, affected by unhealthy conditions of the dominant swamps. This has induced a significant internal migration from northern Italy (especially from the Veneto) to these areas. The slight decrease recorded in the period 1951–1971 should also be noted. After this weak decline, population increased again, thanks to high fertility rates and a considerable internal migration concerning southern regions of Italy.

**Table 4.1** Variation (difference in population density between subsequent decades) in Latium according to elevation classes (mean municipality value).

| Year | Lowlands | Coastal hilly areas | Internal hilly areas | Mountain areas | Latium |
|---|---|---|---|---|---|
| 1871–1881 | 0.24 | 0.10 | 0.08 | 0.01 | 0.04 |
| 1881–1901 | 0.55 | 0.29 | 0.23 | 0.11 | 0.21 |
| 1901–1911 | 0.16 | 0.10 | 0.09 | 0.10 | 0.10 |
| 1911–1921 | 0.17 | 0.17 | 0.06 | 0.03 | 0.06 |
| 1921–1931 | 0.32 | 0.13 | 0.06 | –0.04 | 0.04 |
| 1931–1936 | 1.78 | 0.05 | 0.06 | –0.03 | 0.10 |
| 1936–1951 | 0.68 | 0.21 | 0.09 | –0.01 | 0.09 |
| 1951–1961 | 0.38 | 0.16 | –0.04 | –0.15 | –0.04 |
| 1961–1971 | 0.38 | 0.12 | –0.05 | –0.18 | –0.06 |
| 1971–1981 | 0.21 | 0.17 | 0.07 | –0.07 | 0.04 |
| 1981–1991 | 0.14 | 0.16 | 0.07 | –0.01 | 0.06 |
| 1991–2001 | 0.09 | 0.08 | 0.04 | –0.02 | 0.03 |
| 2001–2011 | 0.07 | 0.05 | 0.01 | –0.02 | 0.01 |
| Mean | 0.42 | 0.14 | 0.06 | –0.02 | 0.05 |

The most marked increase in population density was observed throughout the period in lowlands (mean increase ≈42%) followed by hilly areas close to the sea (mean increase ≈14%). Inner, hilly areas have recorded a slight increase (mean increase ≈6%), while mountain areas showed a slight decrease (mean decrease ≈2%), indicating the major role of the elevation gradient in the context of population displacements. A preliminary analysis on the population density matrix indicates that a multivariate factor analysis is particularly suitable for this kind of dataset. The Kaiser–Meyer–Olkin measure of sampling adequacy and the Bartlett's test of sphericity were adopted. The first one tests whether the partial correlations among variables are small (0.85), whereas the second one tests whether the correlation matrix is an identity matrix ($\chi^2 = 2344$, df = 66, and $p < 0.0001$). Both of them suggest that the model is appropriate to analyse these data. Lastly, the PCA performed on the collected data, extracted the first two axes accounting for about 97% of the total variance, specifically 76% the first axis and 21% the second axis.

Factor loadings on the first two axes illustrated in Figure 4.2 document a temporal path starting in the late 19th century (positively associated with the first axis and negatively with the second) and continuing up to the late

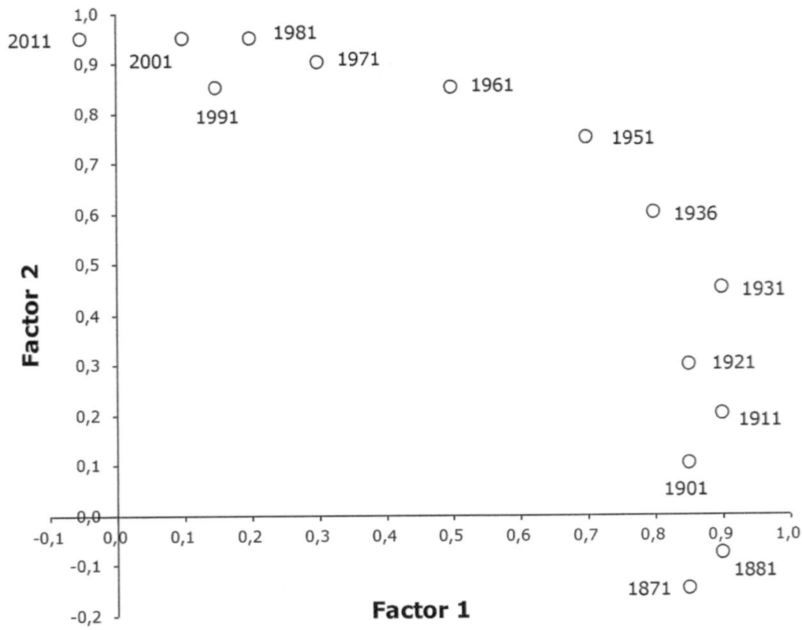


**Figure 4.2** Factor loading plot of population density of Latium from 1871 to 2011.

20th century (showing the opposite behaviour: positively associated with the second axis and negatively with the first). Factor loadings on the first two axes point up population increase recorded during the investigated time window. Visual inspection of factor loadings and scores suggests interpreting the first axis as the urban–rural gradient, while the second axis can represent population displacements along the elevation gradient from lowlands to inner mountain areas.

Results of the *k*-means cluster analysis are finally shown in Figure 4.3. Latium municipalities appear divided in three homogeneous classes, distinguishable by specific demographic dynamics over time. The first cluster, labelled as "increasing" municipalities, contains all the municipalities close to Rome and located along the coastline. They are characterized by three peculiar features: high population pressure, urban sprawl phenomena, and a significant demographic growth in recent years (1971–2001). The profile of the second cluster (labelled as "stable") is more complex. These municipalities include both inner lowland areas and rural hilly zones with medium levels of population density, due to an expansion

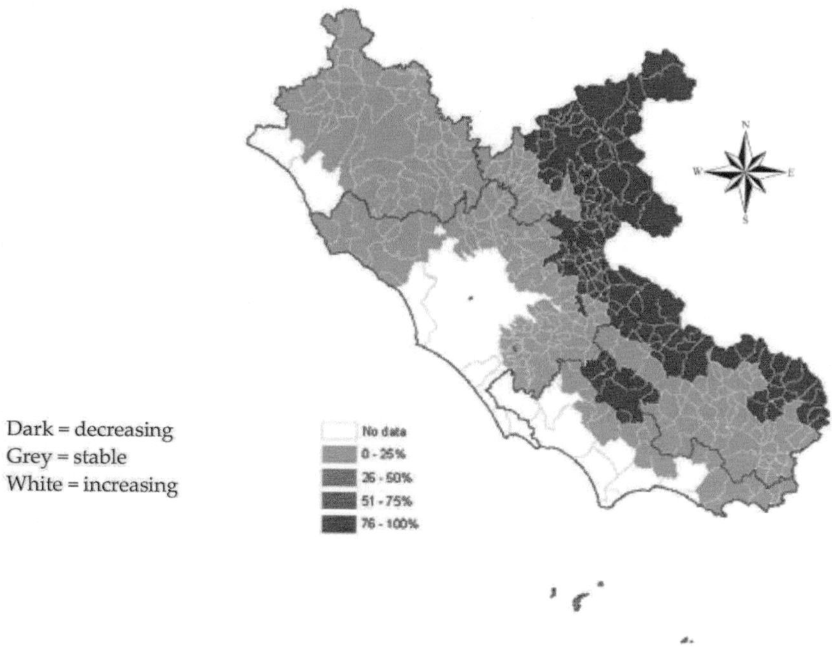


**Figure 4.3** Municipal clusters on the basis of dominant demographic dynamics occurred in Latium region in the period 1871–2011.

phase in the period 1871–1951 and a stable phase in the subsequent time (1951–2011), moderate levels of urban growth, and a tendency to agricultural intensification. The third cluster (labelled as "decreasing") encompasses mountain municipalities located along the Apennines with a typical demographic increase limited to the period 1881–1921 and an almost unstoppable decrease since 1931.

## 4.4 Discussion

Population displacements can be environmentally induced but often are the results of complex socioeconomic mechanisms contributing to create a permanently unstable demographic condition. This can have detrimental repercussions on the natural capital stocks available in that area and subsequently on the life quality of the living population (Garcia Latorre *et al.*, 2001). The multifaceted interactions existing among social, political, economic, environmental, and, recently, also climatic factors involved in a mass migration determine a site-specific demographic dynamics at local, regional, national, and global level (Goria, 2000). In practical terms, migrations are the response to the increasingly widespread imbalance between carrying capacity and population pressure of a given area, looking for a most stable socioecological equilibrium (Knerr, 1998).

Studying in detail population movements is essential to gain knowledge about the primary causes of the demographic phenomena and evaluate the impacts on the natural capital in both physical and economic terms. This way, it is possible to conceive policies directed, on the one side, to mitigate/prevent some socioeconomic side-effects and, on the other side, to preserve/enhance natural resources (Herrmann and Hutchinson, 2005). This effort of conciliating economic development and ecological conservation is particularly delicate in Mediterranean environments, such as the area investigated in this work. Mediterranean ecomosaics along coastline suffer from old and recent waves of degradation (Glenn *et al.*, 1998; Hill *et al.*, 2008; Zdruli, 2014) as a mix of different drivers: demographic boom, seasonal tourism concentration, and agricultural intensification. This provokes high water consumption rates, land take, fires, soil salinization phenomena all contributing to increase the vulnerability to land degradation (Rico-Amoros *et al.*, 2009; Atik *et al.*, 2010; Satriani *et al.*, 2012; Garcia-Ayllon, 2018).

Specifically, Latium recorded a continuous positive demographic trend during the time period 1871–2011. Population movements are mainly directed from inland rural areas to coastal municipalities (Salvati and Zitti, 2005). The adopted multivariate approach testifies the importance of two gradients: urban–rural and elevation. Areas considered economically marginal (i.e., rural countryside and inner mountain areas), on the contrary, show a population decline. This phenomenon can be seen as the long-term effect of the unbalanced population dynamics between disadvantageous internal areas and more productive, lowland, and coastal zones (Tanrivermis, 2003).

This process has induced wide land abandonment phenomena starting since 1950s when people leave marginal areas because of national (industrialization, increased costs of agricultural management, and decreases in agricultural profits) and international (changes in trade regulation between countries) dynamics (Salvati and Zitti, 2005). In the same time, almost everywhere in the Mediterranean areas, population has concentrated mostly in coastal municipalities giving a stimulus to the urban expansion, often at the expense of fertile areas, previously devoted to a profitable agricultural production (Brouwer *et al.*, 1991; Imbrenda *et al.*, 2013; Imbrenda *et al.*, 2021; Coluzzi *et al.*, 2019; Coluzzi *et al.*, 2022; Lanfredi *et al.*, 2022). In the specific case of Latium, this tendency towards urbanization and densification is accompanied by considerable land use/cover changes resulting in degradation of high-productivity soils, moderate increase in fire risk, loss and fragmentation of natural environments and agricultural areas, and salinization phenomena (soil and groundwater; see Mancini *et al.*, 2005; Salvati *et al.*, 2012; Sappa *et al.*, 2019; Bajocco *et al.*, 2020; Bianchini *et al.*, 2021; Nickayin *et al.*, 2021).

## 4.5 Concluding Remarks

Given the considerable importance of demographic dynamics with respect to the implications on environmental/ecological and socioeconomic components, a diachronic approach of the population structural changes is particularly appropriate. This kind of analysis appears further suitable to sound out just Mediterranean environments experiencing a variety of population trajectories from the end of the 19th century up today. In this study, we used effective indicators in the context of a multivariate approach revealing spatial and temporal direction of population in Latium in the period

1871–2011. Findings put in evidence the role assumed by gradients: those existing along urban–rural transitions and along elevation belts.

## References

Agnoletti, M., Errico, A., Santoro, A., Dani, A., & Preti, F. (2019). Terraced landscapes and hydrogeological risk. Effects of land abandonment in Cinque Terre (Italy) during severe rainfall events. Sustainability, 11(1), 235.

Atik, M., Altan, T., & Artar, M. (2010). Land use changes in relation to coastal tourism developments in Turkish Mediterranean. Polish Journal of Environmental Studies, 19(1).

Bajocco, S., Ceccarelli, T., Smiraglia, D., Salvati, L., & Ricotta, C. (2016). Modeling the ecological niche of long-term land use changes: The role of biophysical factors. Ecological Indicators, 60, 231-236.

Bajocco, S., Dragoz, E., Gitas, I., Smiraglia, D., Salvati, L., & Ricotta, C. (2015). Mapping forest fuels through vegetation phenology: The role of coarse-resolution satellite time-series. PloS One, 10(3), e0119811.

Bajocco, S., Ferrara, C., Guglietta, D., & Ricotta, C. (2020). Easy-to-interpret procedure to analyze fire seasonality and the influence of land use in fire occurrence: a case study in central Italy. Fire, 3(3), 46.

Biasi, R., Brunori, E., Smiraglia, D., & Salvati, L. (2015). Linking traditional tree-crop landscapes and agro-biodiversity in Central Italy using a database of typical and traditional products: A multiple risk assessment through a data mining analysis. Biodiversity and Conservation, 24(12), 3009-3031.

Blaikie, P. & Brookfield, H.C. (1987). *Land Degradation and Society*. Methuen, London.

Boccia, L., Capolupo, A., Rigillo, M., & Russo, V. (2020). Terrace abandonment hazards in a Mediterranean cultural landscape. Journal of Hazardous, Toxic, and Radioactive Waste, 24(1), 04019034.

Brouwer, F.B., Thomas, A.J. & Chadwick, M.J. eds. (1991). Land use changes in Europe. In: *Processes of Change, Environmental Transformations and Future Patterns*. Kluwer Academic Publishers, Dordrecht.

Carlucci, M., Chelli, F.M., & Salvati, L. (2018). Toward a new cycle: Short-term population dynamics, gentrification, and re-urbanization of Milan (Italy). Sustainability, 10(9), 3014.

Cecchini, M., Zambon, I., Pontrandolfi, A., Turco, R., Colantoni, A., Mavrakis, A., & Salvati, L. (2019). Urban sprawl and the 'olive' landscape: Sustainable land management for 'crisis' cities. GeoJournal, 84(1), 237-255.

Chelleri, L., Schuetze, T., & Salvati, L. (2015). Integrating resilience with urban sustainability in neglected neighborhoods: Challenges and opportunities of transitioning to decentralized water management in Mexico City. Habitat International, 48, 122-130.

Chianucci, F., Puletti, N., Grotti, M., Bisaglia, C., Giannetti, F., Romano, E., Brambilla, M., Mattioli, W., Cabassi, G., Bajocco, S., Li, L., Chirici, G., Corona, P., & Tattoni, C. (2021). Influence of image pixel resolution on canopy cover estimation in poplar plantations from field, aerial and satellite optical imagery. Annals of Silvicultural Research, 46, 8-13. DOI: 10.12899/asr-2074 2-s2.0-85103753989

Chirici, G., Giannetti, F., McRoberts, R., Travaglini, D., Pecchi, M., Maselli, F., Chiesi, M., & Corona, P. (2020). Wall-to-wall spatial prediction of growing stock volume based on Italian National Forest Inventory plots and remotely sensed data. International Journal of Applied Earth Observation and Geoinformation, 84, 101959. DOI: 10.1016/j.jag.2019.101959 WOS:000501404500018

Chopra, K. & Gulati, S.C. (1997). Environmental degradation and population movements: The role of property rights. Environmental and Resource Economics, 9, 383-408.

Ciommi, M., Chelli, F.M., Carlucci, M., & Salvati, L. (2018). Urban growth and demographic dynamics in southern Europe: Toward a new statistical approach to regional science. Sustainability, 10(8), 2765.

Ciommi, M., Chelli, F.M., & Salvati, L. (2019). Integrating parametric and non-parametric multivariate analysis of urban growth and commuting patterns in a European metropolitan area. Quality & Quantity, 53(2), 957-979.

Coluzzi, R., Bianchini, L., Egidi, G., Cudlin, P., Imbrenda, V., Salvati, L., & Lanfredi, M. (2022). Density matters? Settlement expansion and land degradation in Peri-urban and rural districts of Italy. Environmental Impact Assessment Review, 92, 106703.

Coluzzi, R., D'Emilio, M., Imbrenda, V., Giorgio, G.A., Lanfredi, M., Macchiato, M., Simoniello, T., & Telesca, V. (2019). Investigating climate variability and long-term vegetation activity across heterogeneous Basilicata agroecosystems. Geomatics, Natural Hazards and Risk, 10(1), 168-180.

Corona, P. (2018). Communicating facts, findings and thinking to support evidence-based strategies and decisions. Annals of Silvicultural Research, 42, 1-2. DOI: 10.12899/ASR-1617

Corona, P., Cutini, A., Chiavetta, U., & Paoletti, E. (2016). Forest-food nexus: A topical opportunity for human well-being and silviculture. Annals of Silvicultural Research, 40, 1-10. DOI: 10.12899/ASR-1181.

Cuffaro, N. (2001). *Population, Economic Growth and Agriculture in Less Developed Countries*. Routledge, New York.

Delfanti, L., Colantoni, A., Recanatesi, F., Bencardino, M., Sateriano, A., Zambon, I., & Salvati, L. (2016). Solar plants, environmental degradation and local socioeconomic contexts: A case study in a Mediterranean country. Environmental Impact Assessment Review, 61, 88-93.

Duvernoy, I., Zambon, I., Sateriano, A., & Salvati, L. (2018). Pictures from the other side of the fringe: Urban growth and peri-urban agriculture in a post-industrial city (Toulouse, France). Journal of Rural Studies, 57, 25-35.

Fares, S., Bajocco, S., Salvati, L., Camarretta, N., Dupuis, J.L., Xanthopoulos, G., Guijarro, M., Madrigal, J., Hernando, C., & Corona, P. (2017). Characterizing potential wildland fire fuel in live vegetation in the Mediterranean region. Annals of Forest Science, 74, 1-12. DOI: 10.1007/s13595-016-0599-5

Ferrara, A., Kelly, C., Wilson, G.A., Nolè, A., Mancino, G., Bajocco, S., & Salvati, L. (2016). Shaping the role of 'fast' and 'slow' drivers of change in forest-shrubland socio-ecological systems. Journal of Environmental Management, 169, 155-166.

Garcia-Latorre, J., & Sanchez-Picon, A. (2001). Dealing with aridity: socio-economic structures and environmental changes in an arid Mediterranean region. Land Use Policy, 18, 53-64.

García-Ayllón, S. (2018). Retro-diagnosis methodology for land consumption analysis towards sustainable future scenarios: Application to a Mediterranean coastal area. Journal of Cleaner Production, 195, 1408-1421.

García-Ruiz, J.M., & Lana-Renault, N. (2011). Hydrological and erosive consequences of farmland abandonment in Europe, with special reference to the Mediterranean region–A review. Agriculture, Ecosystems & Environment, 140(3-4), 317-338.

Glenn, E., Stafford Smith, M. & Squires, V. (1998). On our failure to control desertification: Implications for global change issues, and research agenda for the future. Environmental Science & Policy, 1, 71-78.

Goria, A. (2000). Desertification and migration in the Mediterranean: An analytical framework. In: Enne, G. & Zucca, C. (eds.). *Desertification Indicators for the European Mediterranean Region. State of the Art and Possible Methodological Approaches*. pp. 207–223. Agenzia Nazionale per la Protezione dell'Ambiente, Rome.

Herrmann, S.M., & Hutchinson, C.F. (2005). The changing contexts of the desertification debate. Journal of Arid Environments, 63, 538-555.

Hill, J., Stellmes, M., Udelhoven, T., Röder, A., & Sommer, S. (2008). Mediterranean desertification and land degradation: Mapping related land use change syndromes based on satellite observations. Global and Planetary Change, 64(3-4), 146-157.

Imbrenda, V., Coluzzi, R., Lanfredi, M., Loperte, A., Satriani, A., & Simoniello, T. (2018). Analysis of landscape evolution in a vulnerable coastal area under natural and human pressure. Geomatics, Natural Hazards and Risk, 9(1), 1249-1279.

Imbrenda, V., D'Emilio, M., Lanfredi, M., Ragosta, M., & Simoniello, T. (2013). Indicators of land degradation vulnerability due to anthropic factors: Tools for an efficient planning. Sustainable Practices: Concepts, Methodologies, Tools, and Applications, 3, 1400-1413.

Imbrenda, V., Quaranta, G., Salvia, R., Egidi, G., Salvati, L., Prokopovà, M., Coluzzi, R., & Lanfredi, M. (2021). Land degradation and metropolitan expansion in a peri-urban environment. Geomatics, Natural Hazards and Risk, 12(1), 1797-1818.

ISTAT: Circoscrizioni statistiche. Metodi e Norme, Serie C(1), Rome, Italy, (1958).

Knerr, B. (1998). Impact of labour migration on the sustainability of agricultural development in arid regions. In: Clarke, J. & Noin, D. eds. *Population and Environment in Arid Regions* (Man and Biosphere Series 19). UNESCO, New York.

Kosmas, C., Karamesouti, M., Kounalaki, K., Detsis, V., Vassiliou, P., & Salvati, L. (2016). Land degradation and long-term changes in agro-pastoral systems: An empirical analysis of ecological resilience in Asteroussia-Crete (Greece). Catena, 147, 196-204.

Lamonica, G.R., Recchioni, M.C., Chelli, F.M., & Salvati, L. (2020). The efficiency of the cross-entropy method when estimating the technical coefficients of input–output tables. Spatial Economic Analysis, 15(1), 62-91.

Lanfredi, M., Egidi, G., Bianchini, L., & Salvati, L. (2022). One size does not fit all: A tale of polycentric development and land degradation in Italy. Ecological Economics, 192, 107256.

Mancini, L., Formichetti, P., Anselmo, A., Tancioni, L., Marchini, S., & Sorace, A. (2005). Biological quality of running waters in protected areas: The influence of size and land use. Biodiversity & Conservation, 14(2), 351-364.

Mancino, G., Nolè, A., Salvati, L., & Ferrara, A. (2016). In-between forest expansion and cropland decline: A revised USLE model for soil erosion risk under land-use change in a Mediterranean region. Ecological Indicators, 71, 544-550.

Masini, E., Tomao, A., Barbati, A., Corona, P., Serra, P., & Salvati, L. (2019). Urban growth, land-use efficiency and local socioeconomic context: A comparative analysis of 417 Metropolitan regions in Europe. Environmental Management, 63(3), 322-337.

Nickayin, S.S., Salvati, L., Coluzzi, R., Lanfredi, M., Halbac-Cotoara-Zamfir, R., Salvia, R., Quaranta, G., Alhussen, A., & Gaburova, L. (2021). What happens in the city when long-term urban expansion and (un) sustainable fringe development occur: The case study of Rome. ISPRS International Journal of Geo-Information, 10(4), 231.

Nocentini, S., Buttoud, G., Ciancio, O., & Corona, P. (2017). Managing forests in a changing world: The need for a systemic approach. A review. Forest Systems, 26(1), eR01. DOI: 10.5424/fs/2017261-09443

Pender, J.L. (1998). Population growth, agricultural intensification, induced innovation and natural resource sustainability: An application of neoclassical growth theory. Agricultural Economics, 19, 99-112.

Perrin, C., Nougarèdes, B., Sini, L., Branduini, P., & Salvati, L. (2018). Governance changes in Peri-urban farmland protection following decentralisation: A comparison between Montpellier (France) and Rome (Italy). Land Use Policy, 70, 535-546.

Quaranta, G., Salvia, R., Salvati, L., Paola, V.D., Coluzzi, R., Imbrenda, V., & Simoniello, T. (2020). Long-term impacts of grazing management on land degradation in a rural community of Southern Italy: Depopulation matters. Land Degradation & Development, 31(16), 2379-2394.

Recanatesi, F., Clemente, M., Grigoriadis, E., Ranalli, F., Zitti, M., & Salvati, L. (2016). A fifty-year sustainability assessment of Italian agro-forest districts. Sustainability, 8(1), 32.

Rico-Amoros, A.M., Olcina-Cantos, J., & Saurí, D. (2009). Tourist land use patterns and water demand: Evidence from the Western Mediterranean. Land Use Policy, 26(2), 493-501.

Rondeux, J., Bertini, R., Bastrup-Birk, A., Corona, P., Latte, N., McRoberts, R.E., Ståhl, G., Winter, S., & Chirici, G. (2012). Assessing deadwood using harmonized national forest inventory data. Forest Science, 58, 269-283. DOI: 10.5849/forsci.10-057

Salvati, L., & Zitti, M. (2005). Land degradation in the Mediterranean basin: Linking bio physical and economic factors into an ecological perspective. Biota – Journal of Ecology and Biology, 5, 67-77.

Salvati, L., Ferrara, A., & Chelli, F. (2018). Long-term growth and metropolitan spatial structures: An analysis of factors influencing urban patch size under different economic cycles. Geografisk Tidsskrift (Danish Journal of Geography), 118(1), 56-71.

Salvati, L., Gemmiti, R., & Perini, L. (2012). Land degradation in Mediterranean urban areas: An unexplored link with planning? Area, 44(3), 317-325.

Salvati, L., Perini, L., Ceccarelli, T., Zitti, M., & Bajocco, S. (2011). Towards a process-based evaluation of soil vulnerability to degradation: A spatio-temporal approach in Italy. Ecological Indicators, 11(5), 1216-1227.

Salvati, L., Perini, L., Sabbi, A., & Bajocco, S. (2012). Climate aridity and land use changes: A regional scale analysis. Geographical Research, 50(2), 193-203.

Salvati, L., Petitta, M., Ceccarelli, T., Perini, L., Di Battista, F., & Scarascia, M.E.V. (2008). Italy's renewable water resources as estimated on the basis of the monthly water balance. Irrigation and Drainage, 57(5), 507-515.

Salvati, L., & Zitti, M. (2007). Territorial disparities, natural resource distribution, and land degradation: A case study in southern Europe. Geojournal, 70(2), 185-194.

Salvati, L., & Zitti, M. (2009). Substitutability and weighting of ecological and economic indicators: Exploring the importance of various components of a synthetic index. Ecological Economics, 68(4), 1093-1099.

Salvati, L., Zitti, M., & Sateriano, A. (2013). Changes in city vertical profile as an indicator of sprawl: Evidence from a Mediterranean urban region. Habitat International, 38, 119-125.

Sappa, G., De Filippi, F. M., Ferranti, F., & Iacurto, S. (2019). Environmental issues and anthropic pressures in coastal aquifers: A case study in Southern Latium Region. Acque Sotterranee-Italian Journal of Groundwater, 8(1).

Satriani, A., Loperte, A., Imbrenda, V., & Lapenna, V. (2012). Geoelectrical surveys for characterization of the coastal saltwater intrusion in Metapontum forest reserve (Southern Italy). International Journal of Geophysics.

Smiraglia, D., Ceccarelli, T., Bajocco, S., Perini, L., & Salvati, L. (2015). Unraveling landscape complexity: Land use/land cover changes and landscape pattern dynamics (1954–2008) in contrasting Peri-urban and agro-forest regions of northern Italy. Environmental Management, 56(4), 916-932.

Steer, A. (1998). Making development sustainable. Advances in Geo-Ecology, 31, 857-865.

Tanrivermis, H. (2003). Agricultural land use change and sustainable use of land resources in the Mediterranean region of Turkey. Journal of Arid Environments, 54, 553-564.

Tarolli, P., Preti, F., & Romano, N. (2014). Terraced landscapes: From an old best practice to a potential hazard for soil degradation due to land abandonment. Anthropocene, 6, 10-25.

Tomao, A., Quatrini, V., Corona, P., Ferrara, A., Lafortezza, R., & Salvati, L. (2017). Resilient landscapes in Mediterranean urban areas: Understanding factors influencing forest trends. Environmental Research, 156, 1-9.

Zambon, I., Benedetti, A., Ferrara, C., & Salvati, L. (2018). Soil matters? A multivariate analysis of socioeconomic constraints to urban expansion in Mediterranean Europe. Ecological Economics, 146, 173-183.

Zambon, I., Colantoni, A., Carlucci, M., Morrow, N., Sateriano, A., & Salvati, L. (2017). Land quality, sustainable development and environmental degradation in agricultural districts: A computational approach based on entropy indexes. Environmental Impact Assessment Review, 64, 37-46.

Zambon, I., Colantoni, A., & Salvati, L. (2019). Horizontal vs vertical growth: Understanding latent patterns of urban expansion in large metropolitan regions. Science of the Total Environment, 654, 778-785.

Zdruli, P. (2014). Land resources of the Mediterranean: status, pressures, trends and impacts on future regional development. Land Degradation & Development, 25(4), 373-384.

# 5

# Toward a Sustainable Urbanization in Rural Districts? Insights from a Case Study in Central Italy

**Gianluca Egidi[1], Rosanna Salvia[2], Caterina Samela[3], Barbara Ermini[4], Luca Salvati[5], and Leonardo Bianchini[6]**

[1]Department of Agricultural and Forest Science, University of Tuscia, Viterbo, Italy
[2]Department of Mathematics, Computer Science and Economics, University of Basilicata, Potenza, Italy
[3]Institute of Methodologies for Environmental Analysis, National Research Council of Italy (IMAA-CNR), 85050, Tito Scalo, Italy
[4]Department of Economic and Social Sciences, Marche Polytechnic University, Ancona, Italy
[5]Department of Methods and Models for Economics, Territory and Finance (MEMOTEF), Sapienza University of Rome, Rome, Italy

## Abstract

Urban expansion adapts to rural areas, as far as shape and structure, on the basis of the local socio-economic backgrounds and the spatial configuration of forms. Due to urbanization pressure in rural areas, landscapes significantly transformed over time. In traditional peripherical regions of Europe, urban dispersion has become to blend the typical rural landscapes with a new residential settlement matrix based on low-density patches. In Italy, sprawl has been particularly intense, thanks to the lack of compliance with residential planning rules. Tuscia (province of Viterbo, near the city of Rome) is a rural district located in Central Italy and characterized by traditional crops of high agricultural value coexisting with urban dispersion due to "tolerant" urban plans. Therefore, in these areas, urban sprawl phenomena are intrinsically associated with agricultural activities. Our work originates from the assumption that, while the study area is more liveable compared to the neighbourhoods around the city of Rome, urbanization processes in Tuscia are shaped by mixed processes of socio-economic transformations associated with rural population trends that depend on the primary sector. This, in turn, has contributed to preserve a strong naturalness of the local context. Ultimately, Tuscia is probably one of the most pertinent Mediterranean local contexts where spatial dynamics have better adapted to rural communities, influencing the long-term development growth.

## 5.1 Introduction

Landscapes have transformed over centuries in Mediterranean Europe, reflecting the millenary action of rural communities shaping land resources on the basis of environmental constraints and socio-demographic contexts (Salvati *et al.*, 2016; Malek *et al.*, 2018; Varga *et al.*, 2018). These adaptative processes have guaranteed food, energy, and the preservation of traditional lifestyles and building materials, while maintaining a cultural landscape with a wealth of agronomic practices and construction techniques (field works, crop rotation, drainage, and soil tillage) since the Middle Age (Mueller *et al.*, 2021). In the past decades, especially in Southern Europe, urbanization and industrialization have led to crop intensification in the more accessible rural districts and abandonment of marginal areas causing environmental degradation and relevant social problems (Conacher and Sala, 1998; Basso *et al.*, 2000; Satriani *et al.*, 2012; Abu Hammad and Tumeizi, 2012; Imbrenda *et al.*, 2013; Salvati, 2014a; Coluzzi *et al.*, 2019; Quaranta *et al.*, 2020). Since the 1970s, rural landscapes underwent drastic changes under the persistent action of multifaceted socio-economic forces (Meeus, 1993; Salvati, 2014b; De Montis *et al.*, 2017). Urban growth increasingly involved rural areas further away from inner cities (Bruegmann, 2005), giving rise to other transformations including landscape fragmentation, water pollution, land-use homologation, spatial polarization of settlements, alteration of the traditional pattern among land uses, ecosystem degradation, and biodiversity loss (Luck and Wu, 2002; Berling-Wolf and Wu, 2004; Song and Knaap, 2004; Seto and Fragkias, 2005; Couch *et al.*, 2007; DiBari, 2007; Catalàn *et al.*, 2008; Carone *et al.*, 2009; Aguilera *et al.*, 2011; Imbrenda *et al.*, 2014; Salvati *et al.*, 2014b; Serra *et al.*, 2014; D'Emilio *et al.*, 2018; Nickayin *et al.*, 2021). Environmental degradation caused negative consequences in the primary sector with the intense conversion of cropland to discontinuous urban fabric and produced soil sealing, marginalization of rural communities, and loss of sustainable traditional practices (Salvati, 2014b; Lagacherie *et al.*, 2018; Smetanova *et al.*, 2019). These processes are further exacerbated by the pressing threat of climate change that impacts agricultural yields and ecosystem services of natural areas by sharp rise of temperature, reduction of precipitation, changes in precipitation seasonal patterns, increase of number, and intensity of extreme events (Kang *et al.*, 2009; Golinski *et al.*, 2018; Greco *et al.*, 2018; Gupta and Mishra, 2019; Coluzzi *et al.*, 2020; Lanfredi *et al.*, 2020).

Millenary human-landscape interactions (Alphan, 2003) suggest how urbanization may adapt to rural characters with specific shapes and spatial structures, depending on the local context. In Italy, urban sprawl has occurred in various forms and types involving both urban and rural districts (Couch *et al.*, 2007; Salvati and Carlucci, 2016; Gullino *et al.*, 2018) due to its peculiar landscapes that have continuously changed over time. Spatial patterns and settlement forms in Italy were associated with traditional rural landscape transformations oriented toward a more urbanized matrix (Romano *et al.*, 2020). The conversion of natural and agricultural areas to urban uses and the expansion of compact settlements have determined high levels of spatial polarization in population density, settlement forms, incomes, and natural resource distribution leading to significant ecological issues (Salvati and Carlucci, 2011; Salvati *et al.*, 2014a; Fiorini *et al.*, 2019).

In Mediterranean societies, sprawl was initially fuelled by the inclinations of local systems toward illegal housing and land speculation (Weber *et al.*, 2005; Polyzos *et al.*, 2008; Ioannidis *et al.*, 2009; Salvati *et al.*, 2013; Salvati 2014b). Settlement informality and planning deregulation resulted into land appropriation and spatial discontinuous building since the 1970s (Gargiulo, Morelli, and Salvati, 2010). While urban dispersion continued undisturbed during the 1980s and 1990s, its effects are still visible on the land (Catalàn *et al.*, 2008; Chorianopoulos *et al.*, 2010; Fregolent and Vettoretto, 2017). However, local communities with resilient agricultural identity have contained the strong pressure of urban sprawl processes. In these areas, sprawl took place in a milder manner, limiting more severe environmental and landscape damages (Knickel *et al.*, 2018).

The present study investigates urbanization processes, with a special focus on the sprawl phenomena, occurred in Tuscia, a local district located in the heart of Italy. Central Italy is a macro-area where urban and rural zones coexist in prevalently traditional upland landscapes (Bartolini, 2021). Local economic structures are centred on low-income agricultural farms and basic tertiary activities, such as commerce, retail, and public sectors (Salvati, 2014a). Regardless of whether sprawl has occurred, Tuscia can be considered a treasure trove of notable natural and agricultural value, as suggested in Figure 5.1. This statement relies on the assumption that the main constraint to sprawl originates from the socio-economic dimensions of local communities

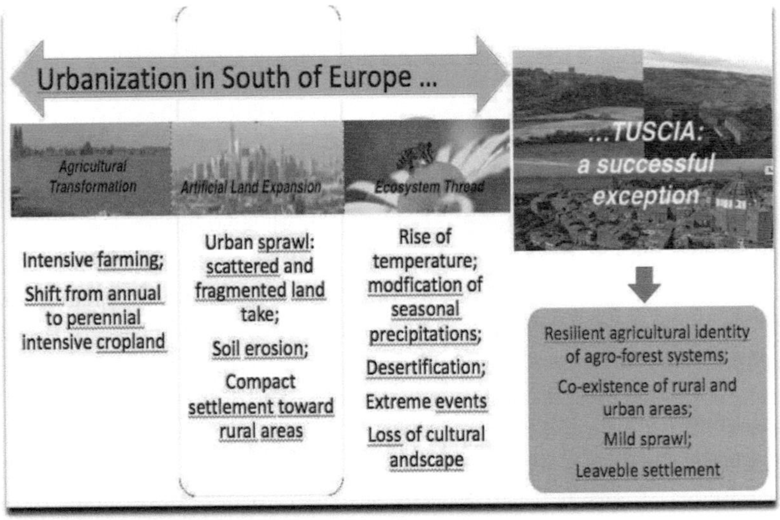


**Figure 5.1** Mediterranean urbanization and Tuscia's peculiarity.

that, while preserving own identity, have also saved a large part of the peculiar Mediterranean landscapes. In these regards, Tuscia is viewed as one of the Italian contexts where urbanization has better adapted to the lifestyle of its inhabitants, and not vice versa.

Although earlier studies have focused on Tuscia agriculture from various points of view (see, e.g., Colantoni *et al.*, 2016), there are few contributions that investigate the relationship between urban and rural areas through economic, social, demographic, and territorial perspectives. Environmental awareness of citizens and entrepreneurs, implementation of suitable policies, and maintaining and management of typical landscapes are demonstrated to be strategic issues for the sustainable development of areas with high landscape-natural value.

## 5.2 Methods

### 5.2.1 Study area

Central Italy, separated from northern Italy by the Apennines, appears as a socially polarized region encompassing urban areas and rural districts with traditional landscapes mainly located in the internal belt (Salvati *et al.*, 2014a). Tuscia coincides with the province of Viterbo (NUTS3 – Nomenclature of territorial units for statistics), located in the north of Latium administrative region (NUTS2), in the heart of Italy (Di Felice *et al.*, 2014) and it is constituted of 60 municipalities surrounding the capital city of Viterbo (about

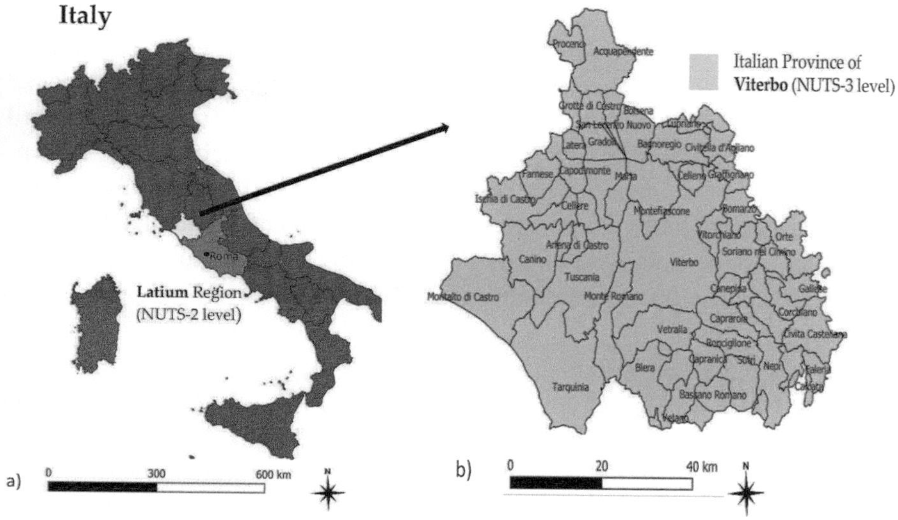


**Figure 5.2** Study area: (a) Tuscia province (Viterbo, NUTS3) located in Latium (NUTS2), in its turn, encapsulated in the Central Italy; (b) municipalities of Tuscia.

70,000 inhabitants). More details are offered in Figure 5.2. This area extends about 3615 km² with a population density of about 88 inhabitants/km².

### 5.2.2 Characteristics of Tuscia landscapes

Tuscia has a vast history of civilization, traditions, and cultural influences, which dates back to the Etruscan period and the Stone Age (Gobattoni *et al.*, 2015). Generally speaking, the "Tuscia" toponym indicated the area initially occupied by the Etruscans, including the northern part of the current Latium region and the southern part of Tuscany. Currently, Tuscia refers largely to the province of Viterbo (Di Felice *et al.*, 2014) being a historical place with a large variety of interesting ingredients: art, archaeology, valuable landscapes, picturesque villages, and uncontaminated nature.

Gross domestic product at market prices per inhabitant of Latium ranked 7th in Italy (about 29,727 euro, from the 2011 Census of Istat – Italian National Institute of Statistics). At the provincial level, employment rate of Viterbo (about 54.4%, as reported by the 2011 Istat census) approached the values found in other provinces of Latium (e.g., Rieti with 54.6% and Rome with 61.1%).

The agro-forestry system is the core element that characterizes Tuscia landscapes, being intimately linked with ancient traditions and including an ecomosaic composed of pastures, grasslands, grain-crops, vineyards, olive groves, and forests. Tuscia economy rests on typical traditional activities connected to the production of food crafts (e.g., cheeses, olive oil, hazelnuts, wine, etc.) and handmade decorative objects (e.g., wood carving, hand-work clay, etc.). The ratio of people employed in agricultural sector in the total workforce was about 2% in Latium region, while for the Tuscia, this value increases up to about 6.4% (2011 Istat Census). Most of Tuscia is occupied by cultivated land, forests, and pastures (corresponding to about 94% of non-sealed land). Forest areas cover about 19% of the overall surface of province. Water heritage includes numerous rivers, torrents, and lakes of volcanic origin (Bolsena). Topography and climate make Tuscia a typical biogeographical area of the central Italy (Colantoni *et al.*, 2016).

### 5.2.3 Early processes of urbanization in Tuscia

Settlement dispersion took place in Tuscia, escaping from the district control of spatial planning while following the rules dictated by the intrinsic agricultural mentality of local communities. Together with low population density, these processes created a more liveable settlement in comparison with the neighbourhoods around Rome, 70 km south of Tuscia. According to Colantoni *et al.* (2015), Rome is a semi-compact and dense Mediterranean city experiencing sprawl and structural changes of its peri-urban landscapes. The main landscape transformations occurred include: (i) mixed developments of settlement densification and dispersion through a progressively degraded and fragmented agricultural landscape, (ii) reduction of the semi-natural land into woodlands covering marginal areas, and (iii) an increasing mixité of traditional rural landscapes with human settlements (Salvati, 2014b). Tuscia low-density settlements were placed in a less chaotic and scattered structure and the majority of buildings had a regular permit for rural construction. Until recently, this title could be obtained rather simply: the possession of minimum rural lots and being agricultural entrepreneurs were essential criteria regardless of the personal income derived from such activity. In this way, a sort of legalization of land consumption was allowed in these contexts. However, rural areas were managed over time to produce a continuous income for many families (De Rosa *et al.*, 2019). Despite having experienced some informal practices, local administrations of Tuscia have kept measures to protect agricultural activities and the local workforce. These policies made possible to better preserve the countryside, avoiding further urbanization.

### 5.2.4 Data analysis

The investigation was performed using statistical techniques and geographical information system (GIS). Municipalities, as unit of analysis, are considered a

**Table 5.1** List of the indicators used in the present work.

| Acronym | Name | Unit of measure | Data source |
|---|---|---|---|
| DEM51-11 | Demographic Variation 1951-2011 | % | Istat - census |
| DEM51-81 | Demographic Variation 1951-1981 | % | " |
| DEM81-11 | Demographic Variation 1981-2011 | % | " |
| DEM01-11 | Demographic Variation 2001-2011 | % | " |
| DEM11-15 | Population Classes 11-15 | % | " |
| DEN | Population Density | inhabitants/$km^2$ | " |
| BUID50 | Buildings Constructed until 1950 / Total Buildings | % | " |
| BUID80 | Buildings Constructed until 1980 / Total Buildings | % | " |
| BUIREC | Buildings recently constructed from 1981 / Total Buildings | % | " |
| AGE | Elderly Index | Index | " |
| MAGE | Medium Age Index | Index | " |
| HS | Household Size (medium) | population/families | " |
| UNIFAM | Family from a Component / Population | % | " |
| EAGR | Employed in Agriculture / Total Employees | % | " |
| ECRED | Employed in Credit and Insurance and Business Services / Total Employees | % | " |
| EMPL | Total Employment Rate | % | " |
| COMM | People living outside the municipalities / Population | % | " |
| LMOB | Long Mobility/Population | Index | " |
| AINC | Average Income Per Capita | € | Istat |
| DIST-VT | Distance from Viterbo | km | " |
| DIST-ROME | Distance from Rome | km | " |

**Table 5.1** Continued

| Acronym | Name | Unit of measure | Data source |
|---|---|---|---|
| SC | Soil Consumption | hectares | |
| NAT | Natural Areas with A Very Low Density / Surface Area | % | Elaboration on Ispra - sinanet. isprambiente web site |
| LDA | Urbanized Areas with Low Density / Surface Area | % | |
| HDA | Artificial Areas with High Density of Urbanization / Surface Area | % | |
| ART | Strictly Artificial Surface / Total Landscape | hectares | |
| RAA | Anthropogenic Area / Total Landscape | % | |
| CROP** | Crops / Surface Area | % | Ispra - elaboration on CLC 2012 (land-use) |
| WOOD*** | Forests / Surface Area | % | |
| WATE | Water System / Surface Area | % | |
| ICROP | Intensive Crops / Surface Area | % | |
| CRS | Residential Areas in Continuous Fabric / Surface Area | % | |
| DRS | Residential Areas Discontinuous tissue / Surface Area | % | |
| CHA1-1 | Conversion from 1 a 1 from 2000 / surface area | % | Ispra - elaboration on land use changes |
| CHA2-1 | Conversion from 2 a 1 from 2000 / surface area | % | CLC |
| CHA2-2 | Conversion from 2 a 2 from 2000 / surface area | % | " |
| CHA2-3 | Conversion from 2 a 3 from 2000 / surface area | % | " |
| CHA2-5 | Conversion from 2 a 5 from 2000 / surface area | % | " |

**Table 5.1** Continued

| Acronym | Name | Unit of measure | Data source |
|---|---|---|---|
| CHA3-1 | Conversion from 3 a 1 from 2000 / surface area | % | " |
| CHA3-2 | Conversion from 3 a 2 from 2000 / surface area | % | " |
| CHA3-3 | Conversion from 3 a 3 from 2000 / surface area | % | " |
| RSS0 | Soil Sealing0 / Surface Area | hectares | EEA - elaboration on soil sealing map |
| RSS1-20 | Soil Sealing1-2 0/ Surface Area | hectares | |
| RSS21-40 | Soil Sealing21-40 / Surface Area | hectares | |
| RSS41-60 | Soil Sealing41-60 / Surface Area | hectares | |
| RSS61-100 | Soil Sealing61-100 / Surface Area | hectares | |
| TEXP | Total Expenditures / Surface Area | € / kmq | Elaboration of the database of the Italian Ministry of Interior |
| MEXP | Mobility Expenses / Surface Area | € / kmq | |
| LEXP | Land and Environment Management Expenses / Surface Area | € / kmq | |
| YPRG | Year of the most recent municipal planning instrument | year | Latium Region's report |

** Including different types of crops (areas with sparse vegetation, predominantly occupied by crops with the presence of important natural areas, extensive crops, temporary crops, and olive groves), excluding intensive crop.

*** Including different types of forest (wooded areas and shrub evolving, woods with a prevalence of chestnut, and forests of beech prevalence or prevalence of Mediterranean pine trees).

relevant spatial domain for investigation of land-use changes, environmental reporting, and sustainable management of rural land (Salvati and Zitti, 2009).

A total of 49 indicators have been developed from a collection of variables regarding each of the 60 municipalities belonging to the Tuscia district (Table 5.1). Most of them were extracted from the Italian National

Census of Population and Buildings (available at https://www4.istat.it/it/censimenti-permanenti/) and mainly referred to the last survey carried out in 2011. Indicators allow a detailed socio-economic characterization of municipalities. A work carried out by Institute for Environmental Protection and Research (ISPRA) was also considered (Munafò *et al.*, 2015a) because it adopts a set of indicators able to investigate how and where soil consumption affects the local context in Italy, providing a national map of soil consumption (2012).

The Corine Land Cover (CLC) map, provided by ISPRA, was used to analyse land-use structure of Tuscia. CLC was classified into the following classes: (i) residential areas in continuous fabric; (ii) residential areas in discontinuous fabric; (iii) anthropogenic area (excluding residential areas); (iv) cropland; (v) intensive cropland; (vi) forests; (vii) water system (including sea, lakes, and rivers). Following the ISPRA approach, the present work considered also land-use change maps (2000–2006 and 2006–2012). The indicators elaborated in this way are able to estimate how areas of each land-use class have been converted to another use since 2000, with the aim to evaluate land-use transformations in Tuscia. Territorial transformation indicates the conversion from the original state of land use to the final use, by grouping the detailed classes into macro-categories from 1 to 5: (1) urban areas; (2) croplands; (3) forests; (4) wetlands; (5) water system.

Another map adopted to validate the reliability of the CLC in Tuscia concerns the spatial distribution of built-up and non-built-up areas, including the continuous degree of soil sealing expressed in percentage terms (from 0 to 100) with an aggregated spatial resolution (100 m × 100 m grid). The link between sprawl and soil sealing is assumed to be strong and the debate around the assessment of soil sealing through effective monitoring schemes is particularly urgent. In this ambit, Italy represents a paradigmatic case where there is an escalation of land conversion to urban uses. Five classes of soil sealing were identified: the first category corresponds to "0%", for natural covers; classes "1%–20%" and "21%–40%" for areas with low density settlement; class "41%–60%" corresponding to settlements with medium density; and lastly the class showing the most dense and compact settlement conditions ("61%–100%").

Data regarding expenses of municipalities were also collected from the Italian Ministry of Interior. Being freely accessible from the website, this database ensures transparency of the institutional activity. In this way, annual final balance sheets (2012) were analysed per total charges, focusing on expenses of mobility, territorial, as well as environmental management. Territorial objectives and strategies, as defined by each Italian region, form a guideline to the provincial and municipal policies and their respective programs and sector plans. The Master Plan (PRG) is the predominant spatial instrument at the municipal scale. The Latium region regularly checks the general situation concerning urban planning. From the analysis of such report, we adopted another variable referring to the year of approval (or update) of the municipal Master Plan.

All these data have been translated into quantitative indicators with the purpose to profile the local background context. Correlations were run to verify whether there is a relationship between two indicators, demonstrating their inclination to change together. A non-hierarchical cluster analysis (using standard *k*-means computation strategy) was performed to classify municipalities into homogeneous spatial groups using Euclidean distances and a "paired with group" amalgamation algorithm. Maps of indicators and the spatial analysis in this study allowed detection of the most relevant contexts, enabling an intrinsic validation of the empirical results (Salvati *et al.*, 2011).

## 5.3 Results

In Figures 5.3(a) and (b), the CLC2012 shows the prevalence of rural land uses. Rural activities are primarily concerned with extensive agricultural systems, while intensive crops are located near the sea, where the largest greenhouse cultivations are settled. Soil sealing and soil consumption maps (Figures 5.3(c) and (e)) point out that the largest concentration of populated areas (soil sealing degree > 60%) and density (Figure 5.3(d)) is close to agricultural areas (Figure 5.3(f)), hosting intensive farming systems and discontinuous residential settlements. Viterbo municipality assumed a predominant artificial configuration with medium–high population density and high percentage of soil sealing compared with the total municipal area.

Demographic changes help to track evolution of settlement in Tuscia districts. Figure 5.4(a) shows how, since the 1950s, the largest population change concentrated around Viterbo; later, Viterbo has recorded a more contained demographic change. A notable demographic escalation characterized the areas surrounding Viterbo and especially coastal areas including the municipalities of Tarquinia and Montalto di Castro (Figure 5.4(b)). Second, home development is responsible for such dynamics. In more recent years

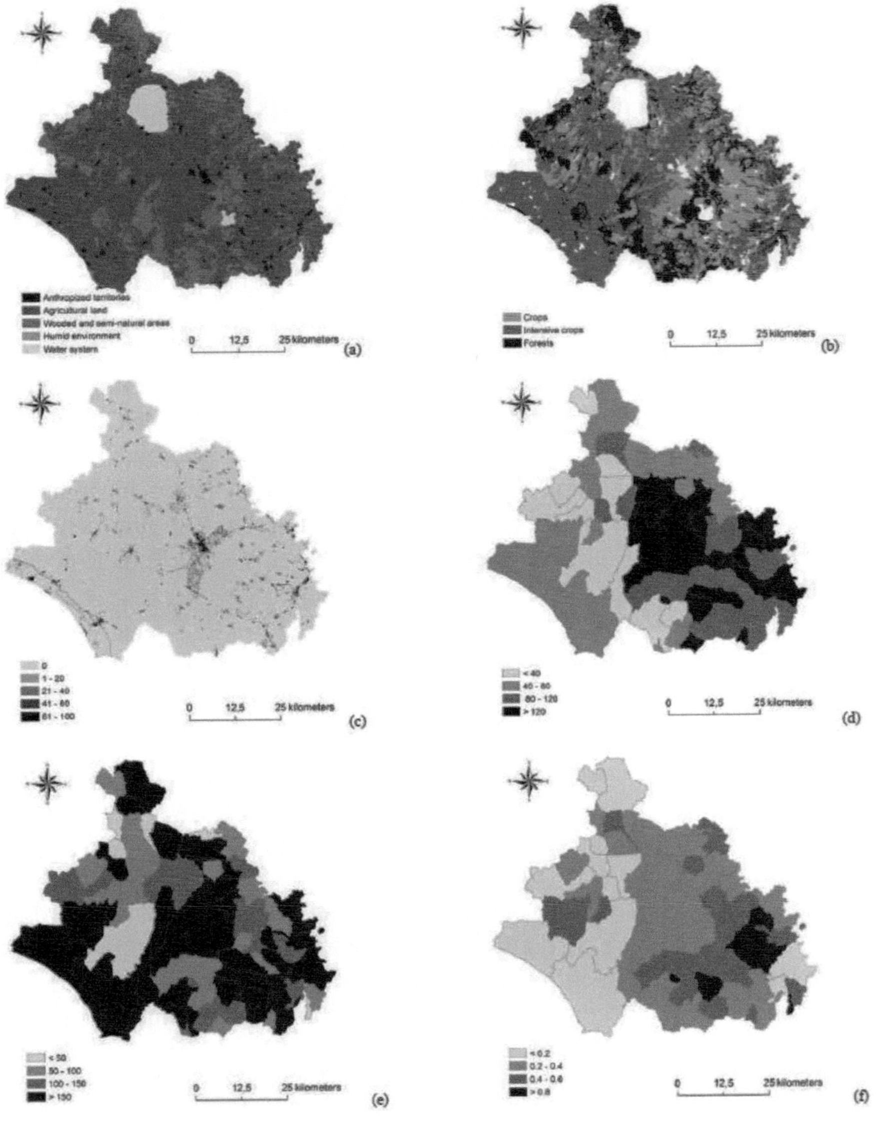


**Figure 5.3.** (a) Analysis of landscape structure in the study area: CLC classification – level 1 (2012). (b) Main categories of the rural environments. (c) Soil sealing. (d) Population density. (e) Soil consumption. (f) Percentage of agricultural land (excluding intensive crops) on the surface area.

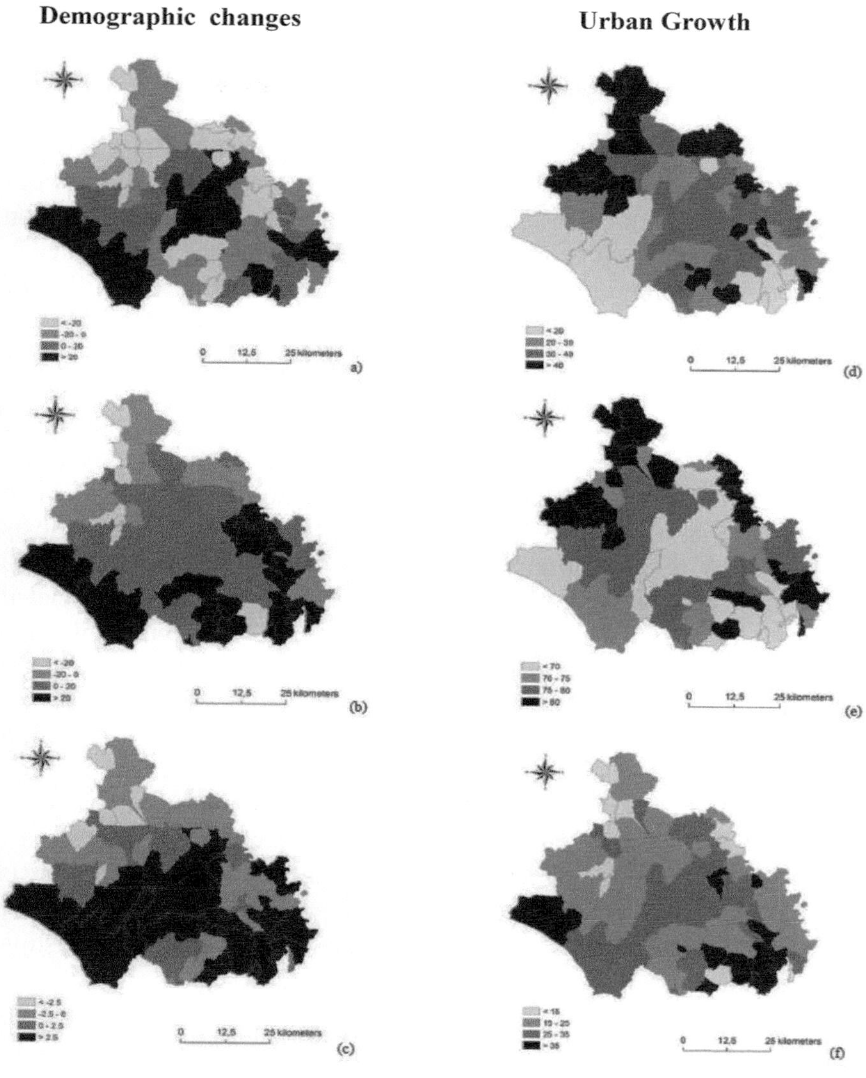


**Figure 5.4** Population changes from 1951 to 2015 and urban growth. Maps correspond to: (a) DEM51-11; (b) DEM51-81; (c) DEM11-15; (d) BUID50; (e) BUID80; (f) BUIREC.

(Figure 5.4(c)), population grew essentially around the city of Viterbo and in coastal municipalities.

Despite demographic and urban change, Tuscia substantially maintained its rural connotation over time.

Results of a correlation analysis clarified the relationship between the selected indicators and help to assess how socio-economic forces have fuelled urban dispersion in Tuscia. Table 5.2 illustrates peri-urban correlation coefficients regarding population density (DEN), elderly index (AGE), percentage of discontinuous residential areas in total municipal area (DRS), and the percentage share of land converted from rural areas to urban fabric since 2000 in total municipal area (CHA2-1). Population density (DEN) is positively correlated with: (a) percentage of new buildings (since 1981) in total building stock; (b) family size; (c) degree of commuting; (d) percentage of total municipal expenses; (e) expenditures relating to the road network and land management of environment in total municipal area (2011); (f) share of low-density urban areas in total landscape; (g) anthropogenic contexts characterized by residential areas with degree of soil sealing between 1% and 60% and by the presence of relict cropland. DEN is negatively correlated with: (a) percentage of buildings constructed until 1980 in total buildings; (b) single-family households in total population; (c) employed in the primary sector in total employed; (d) elderly and middle age index; (e) predominantly natural areas with 0% of soil sealing rate in total municipal area. Correlation demonstrates that sprawl has taken place in Tuscia with a lesser pace in those high natural ecomosaics where agriculture was able to contain an excessive land consumption. A high value of the ageing index and buildings activity, dating back to the 1980s, characterizes such rural contexts. Ageing index is also associated with very low values of demographic change, average per capita income, employment rate, individual–family households (consisting of elderly widowers), and employees in the agricultural sector.

**Table 5.2** Correlation analysis between the selected indicators (population density "DEN", elderly index "AGE", discontinuous residential areas on total municipal area "DRS", and land-use conversion from rural to urban areas from 2000 on total municipal area "CHA2-1") and context variables (significant coefficients > |0.5|).

| **Variables** | **DEN** | **AGE** | **DRS** | **CHA2-1** |
|---|---|---|---|---|
| DEM51-11 | 0.6 | –0.7 | 0.6 | 0.6 |
| DEM81-11 | 0.6 | | | |
| DEM01-11 | 0.6 | –0.7 | 0.7 | 0.5 |

**Table 5.2** Continued

| Variables | DEN | AGE | DRS | CHA2-1 |
|---|---|---|---|---|
| DEM11-15 | 0.6 | –0.7 | 0.6 | 0.5 |
| DIST-ROME | –0.6 | 0.7 | | |
| DEN | 1.0 | –0.6 | 0.8 | |
| BUID50 | | 0.7 | | |
| BUID80 | –0.6 | 0.6 | –0.7 | |
| BUIREC | 0.6 | –0.6 | 0.7 | |
| HS | 0.6 | –0.7 | 0.6 | |
| UNIFAM | –0.6 | 0.6 | | |
| EAGR | –0.5 | 0.7 | | |
| ECRED | | –0.5 | | |
| EMPL | | –0.5 | | |
| AGE | –0.6 | 1.0 | –0.5 | |
| MAGE | –0.6 | 1.0 | –0.6 | |
| COMM | 0.5 | –0.9 | 0.5 | |
| AINC | | –0.6 | | |
| TEXP | 0.9 | –0.5 | 0.8 | |
| MEXP | 0.6 | | 0.5 | |
| LEXP | 0.9 | –0.5 | 0.8 | |
| NAT | –0.7 | 0.5 | –0.7 | |
| LDA | 0.7 | –0.5 | 0.7 | |
| ART | 0.8 | –0.5 | 0.8 | 0.8 |
| SS0 | –0.8 | | –0.7 | –0.6 |
| RSS1-20 | 0.7 | | 0.7 | 0.5 |
| RSS21-40 | 0.8 | | 0.7 | 0.7 |
| RSS41-60 | 0.7 | –0.5 | 0.7 | |
| RAA | 0.5 | | | 0.9 |
| CROP | 0.5 | | | |
| DRS | 0.8 | –0.5 | 1.0 | |
| CHA1-1 | 0.5 | | | 1.0 |
| CHA2-1 | | | | 1.0 |

Typical contexts of sprawl with medium–low density were concentrated in the southern part of Tuscia – close to Rome – and along the coast. The indicator "DRS", referring to "residential areas discontinuous tissue/surface area", is positively correlated with: (a) demographic change; (b) population density; (c) percentage of recent buildings since 1981 and commuters; (d) average households size; (e) the municipal expenditures (and intermediate levels of soil sealing, i.e., between 1% and 60%).

As expected, correlation indicates a modest change in terms of agricultural areas converted to urban land uses since 2000, compared to other land-use changes. The conversion from agricultural areas to urban fabric has involved large part of the municipalities, indicating the presence of land consumption processes. The municipalities of Capranica, Montalto di Castro, Monterosi, Orte, Proceno, and Viterbo have subjected to a re-functionalization of built-up areas. The conversion from agricultural to urban areas is observed mainly in municipalities having medium–high population density accompanied by a great demographic change and soil sealing variations higher than 0. These findings highlight that, generally, the most impacting change from natural environments to urban land uses rarely have occurred in Tuscia impeding the formation of highly fragmented patches.

Wider demographic changes concentrate in those dispersed areas characterized by a soil sealing rate between 1% and 60%. In any case, it is strongly connected to the "density" factor, meaning that there is not excessive dispersion of urban fabric in the surrounding countryside. Here, a large part of young people lives and moves around the area. Generally, they have medium–high income levels, preferring to live close to more dense and compact urban areas. As correlation shows, Tuscia is one of the few Italian cases where most of the soil consumption phenomena are concentrated in urban contexts having medium–high density, resulting in an overall limited land consumption.

The non-hierarchical cluster analysis (using the standard *k*-means technique), carried out to group municipalities into homogeneous spatial clusters by using the Euclidean distance, produces three clusters with peculiar territorial features (Figure 5.5). Viterbo, Ronciglione, and San Lorenzo Nuovo form cluster 1 (black-coloured areas) composed of compact and dense settlements. Prevalently located in the southern part of the province, municipalities such as Vetralla and Caprarola with medium density and surrounded by countryside constitute cluster 2 (dark grey in the legend). Natural areas are predominant in cluster 3 (light grey in the legend), composed of landscapes of high natural value, with occasional low-density settlements (e.g., Acquapendente, Bagnoregio, and Barbarano Romano municipalities).

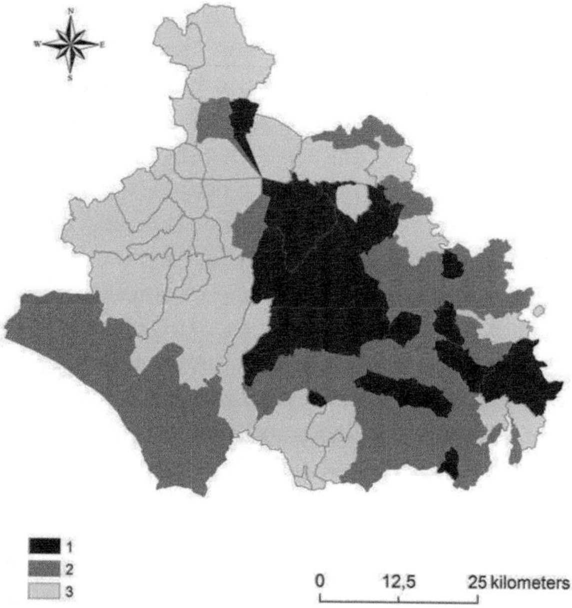


**Figure 5.5** The three different clusters identified in Tuscia through the *k*-means technique.

Subsequently, a hierarchical cluster with similarity index (Euclidean), carried out with the algorithm "paired with group", identifies four groups (Figure 5.6), which substantially confirms the results obtained from the *k*-means analysis. The cluster including the city of Viterbo has peculiar features consisting principally in a compact spatial configuration associated with high values of positive demographic changes (especially in recent years), employment rate, and income per capita. Families of this cluster are rather large, including mainly young people and commuters. As regards urban settlements, they have undergone a notable decline until the 1980s, followed by a significant increase in recent years. This urban expansion is testified by the performance of this cluster reaching the highest provincial values of soil sealing rate, percentage of rural areas converted into urban surface, and soil consumption rate. The municipalities populating this cluster are surrounded by typical rural landscapes encompassing extensive crops (e.g., olive grove and hazelnut) reaching a medium–low level of rural quality as most of other Tuscia municipalities. This environmental condition translates into a sharp increase of expenditures that impact municipal budgets especially for land management issues (e.g., Burchell *et al.*, 2005). A specific group (cluster 2) stands out in the hierarchical clusters with similarity index. The municipalities of Blera, Vejano, Canino, Ischia di Castro, and Arlena di

Castro appear to have specific features. Even though this cluster 2 is similar to cluster 4 (according to the *k*-means procedure, the two clusters belong to the same group), the municipalities belonging to it are distinguishable by their high naturalness and low population density. The socio-demographic profile of these municipalities shows a continuous negative demographic trend and a dominant presence of older and low-income people, which are rather static in the area with a consequent low employment rate despite a mighty component of the workforce in the agricultural sector. This socio-economic structure has preserved rural areas, strongly limiting land consumption rates which are the lowest within the study area. In this cluster, the built-up expansion has stopped in the 1980s, in favour of the agricultural sector, mainly in the intensive form. These socio-economic and landscape features have thus a low impact on municipal budgets. Cluster 3 includes most of the municipalities falling in the southern part of Tuscia with medium–high values of land consumption certifying the presence of urban sprawl phenomena. The real estate expansion has grown over time, especially along the coast where tourism has shown an important escalation in the last decades. Here, urban dispersion could be the consequence of a collective behaviour having effective consequences in socio-economic terms. This cluster, in fact, shows: a steadily demographic gain especially in recent years, a strong territorial mobility, medium–high employment rates, and a notable incidence of young people and large families, all elements confirming the presence of an active population. Cluster 3 has a spatial configuration where medium-density settlements and rural landscapes are mixed and ranging from intensive farming in the west to extensive crops in the east. Lastly, cluster 4 includes the most natural environments, with medium–low density settlements reflecting a low rate of land consumption and limited transformations from rural areas to urban land uses. In fact, large part of cluster 4 is occupied by agricultural uses with a dominance of intensive crops with consequent environmental impacts. From a socio-economic point of view, this cluster is part of the economically disadvantaged and marginal areas marked by negative demographic rates, lower incomes, and employment rates, where socio-economic inequalities and environmental issues are topical (Salvati *et al.*, 2014a).

The presence of four clusters corroborates the hypothesis that land policies have produced different repercussions on the land systems. Results of the present study have highlighted a general status of inertia of the Master Plan (PRG) in the whole Tuscia. The municipalities with the highest population density and land consumption rate (specifically those belonging to the cluster including the city of Viterbo) have a dated PRG. A few exceptions

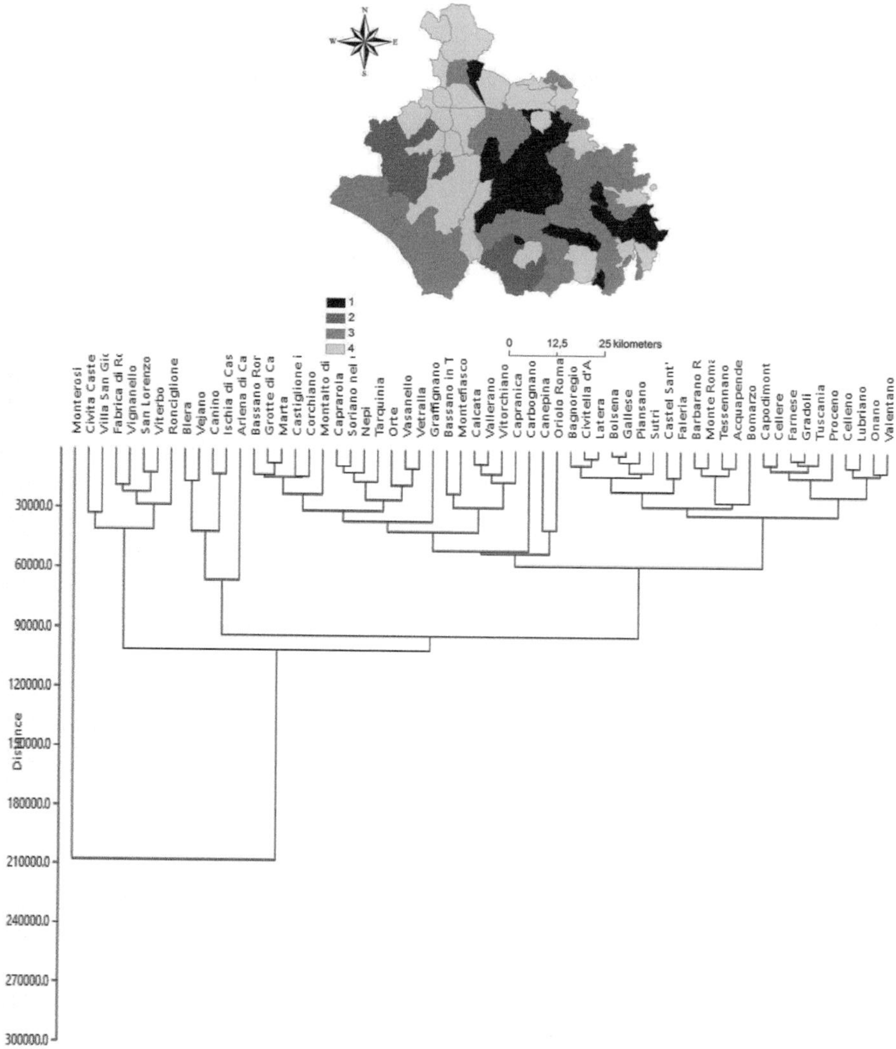


**Figure 5.6** Results of a hierarchical clustering to group municipalities in Tuscia based on socio-economic indicators.

are the municipalities of Soriano nel Cimino, Proceno, and Acquapendente (belonging to other clusters) which recently adopted (from 2010 to present) an adjustment of such instruments (known as "variante"). When Master Plans are not regularly updated, a passive attitude emerges, leading to a defective management of landscape. At municipal scale, this is confirmed by the observed negative correlation (–0.5) between the year of the most recent Master Plan (or PRG) and SC and HDA indicators (i.e., the municipalities with a more dated PRG have recorded greater soil consumption and high-density urbanization rates).

## 5.4 Discussion

The Mediterranean region is a representative example of the millenary interplay between natural ecosystems and anthropic activities (Alphan, 2003; Ferrara *et al.*, 2014; Zimmerer *et al.*, 2020). Since the 1970s, a sequence of various phenomena (e.g., urbanization, land abandonment, crop intensification, soil degradation, and economic dynamics) have taken place in southern Europe determining considerable land-use changes (Paul and Tonts, 2005; Catalàn, *et al.*, 2008; Nolè *et al.*, 2009; Chorianopoulos *et al.*, 2010; Ferrara *et al.*, 2014; Scherer *et al.* 2018). Driving forces, such as urban sprawl (Cakir *et al.*, 2008) and agricultural intensification (Alphan, 2003), have deeply transformed traditional landscapes and induced a more fragmented spatial configuration especially in the most vulnerable rural and peri-urban areas (Marraccini *et al.*, 2015; Salvati *et al.*, 2016; Gariano *et al.*, 2018). The collective costs, linked mainly to urban dispersion processes, have already been studied (e.g., Burchell *et al.*, 2005; Polidoro *et al.*, 2012; Solly *et al.*, 2020). Municipal budgets have paid the price for the occurrence of these phenomena. In particular, higher expenditures are concentrated in the medium–high density contexts with a large presence of sprawled residential areas (Varela-Candamio *et al.*, 2019). The advent of these phenomena has generated a series of reflections ranging from the social–economic (e.g., social segregation and polarization, economic marginality, etc.) to ecological–environmental issues (e.g., land degradation) that affected disadvantaged rural regions of southern Europe (Rubio and Bochet, 1998; Iosifides and Politidis, 2005; Wilson and Juntti, 2005; Salvati and Carlucci, 2011; Salvati *et al.*, 2014a; Imbrenda *et al.*, 2018; Næss *et al.*, 2020). More specifically, Tuscia appears only partially affected by these processes because urban dispersion is confined to areas with medium–high urban density. This land cover dynamics is greatly influenced by the strong local sense of place. This way, natural, semi-natural, and agricultural areas have remained mostly unchanged over time (Palazzo and Aristone, 2017). The municipality of Viterbo constitutes a paradigmatic exemplification of what said above being one of the few Italian cities having a low percentage of land consumption in recent years. The presence of a consolidated rural setting has allowed to limit an excessive soil consumption over time, maintaining the traditional landscape arrangement and the same socio-economic profile (Lopez-Goyburu and García-Montero, 2018).

The present work explores an area poorly investigated in terms of urban–rural gradient evolution. Crop intensity and consequent phenomena of soil compaction affect the study area, whereas it appears as moderately sensitive to urban sprawl, soil sealing, and other environmental issues (e.g., salinization and fire risk) (Salvati *et al.*, 2011). The preservation of rural-cultural landscapes in peri-urban contexts can be a complex task due to the interplay of various factors, such as biophysical, social, economic, cultural, and political elements (García-Ruiz *et al.*, 2020). Peri-urban areas should be treated as rural landscapes, to effectively protect them in the long-term through an ecologically feasible approach, encompassing a mix of measures consisting in containing human pressure and involving social tools relying on the existence of eco-conscious local communities (Salvati *et al.*, 2016; Kreiling *et al.*, 2018; van Vliet *et al.*, 2020). This has happened in Tuscia: land consumption has prevalently affected denser urban areas, leaving larger spaces to the agricultural land uses, limiting effectively the presence of informal dispersion processes. Results reveal that municipalities with updated Master Plans have suffered less damage from land consumption processes. A continuous update of urban planning instruments would allow to avoid potential damaging impacts to the environment and to the built-up heritage, to ensure effective land maintenance and conservation, and to verify compliance with the guidelines dictated by higher institutional levels. The increasing need to implement strategic measures for a more sustainable world imposes a deep reflection on the role of planning tools (Di Ludovico and D'Ascanio, 2019; Shaw *et al.*, 2020; Le Bivic and Melot, 2020). In this context, the General Regional Master Plan (GRMP) of the Latium region requires the achievement of a set of specific objectives at lower territorial levels, such as the promotion of the local identity, the containment of soil sealing and urban sprawl phenomena, and the safeguard of agro-forestry landscapes. According to the GRMP, more specific objectives should be included and adapted to each local context (through the adoption of municipal Master Plan). In fact, by analysing the wide variability of socio-ecological profiles at municipal scale within the Latium region, there are different conditions, such as the city of Rome with its complex metropolitan area and most of the municipalities of Tuscia with a strong rural vocation, where agriculture represents a real added value for the territory. Given this variegated picture, the rebalancing of urban, peri-urban, and rural areas into a new urban-to-rural "sustainable continuum" is essential (Salvati *et al.*, 2016). The focus should be on agricultural activities, which should increase their organization levels particularly in the peri-urban contexts, to reach a better ecological equilibrium including also the neighbouring urban systems and to strengthen their sustainability (Silvestri and Bonari, 2010; La Rosa *et al.*, 2014; Spyra *et al.*, 2020). For example, the development of circular economy in agriculture can limit and control farming wastes, utilizing them as a sustainable strategy (Jun and Xiang, 2011; Secondi *et al.*, 2019; Poponi *et al.*, 2021). Waste from agricultural and food industry can be used as biomass (e.g., biochar), giving energy to productive processes (Pan *et al.*, 2015; Zabaniotou *et al.*, 2015; Colantoni *et al.*, 2016). A sustainable cycle of economy is the way to achieve an actual development involving environment and economy (Jun and Xiang, 2011; Statuto *et al.*, 2019; Muscio and Sisto, 2020). This kind of economy should rely on the awareness that local resources can represent a viable opportunity for the future. For example, the establishment of agro-energy districts can be a standpoint of sustainable development because they can create a network of actors and stakeholders, promote the reuse of waste, natural resources, and other clean energy, increase local sense of place, enhance quality and protection of landscapes, and ensure the maintenance of local businesses (e.g., Shamsuzzoha *et al.*, 2012; Tziolas *et al.*, 2016). This work and similar analysis aimed at identifying other vulnerabilities of the investigated socio-ecological systems, at spatial scales compatible with regional/local planning (Salvati *et al.*, 2016), can help to devise and implement sustainable land management strategies. In the case of Tuscia, these strategies will preserve the high natural value of its rural ecomosaics.

## 5.5 Concluding Remarks

The present paper explores a case study in Central Italy allowing a comparative investigation of its local identity, traditional economic activities, and spatial processes. Urbanization processes in Tuscia have been intimately linked with the primary sector acting as an indirect factor of urban containment. Tuscia, therefore, represents a unique context where socio-economic components, reinforced by local traditional and cultural beliefs, have characterized and preserved its natural landscapes until today. This rural connotation stimulates a debate on future management of traditional cropping systems vis-à-vis urbanization. The area appears as a paradigmatic territorial model that combines traditional activities and progress through strategies that include a variety of sustainable policies such as the establishment and consolidation of agro-energy districts.

## References

Abu Hammad, A. and Tumeizi, A. (2012). Land degradation: socioeconomic and environmental causes and consequences in the eastern Mediterranean. Land Degradation & Development, 23, 216-226.

Aguilera, F., Valenzuela, L.M. and Botequilha-Leitão A. (2011). Landscape metrics in the analysis of urban land use patterns: A case study in a Spanish metropolitan area. Landscape and Urban Planning, 99, 226-238.

Alphan, H. (2003). Land use change and urbanisation of Adana, Turkey. Land Degradation & Development, 14, 575-586.

Bartolini, F. (2021). Back to a future civilization: cities and countryside in the 'Third Italy'. Urban History, 48(1), 108-124.

Basso, F., Bove E., Dumontet S., Ferrara A., Pisante M., Quaranta G. *et al.* (2000). Evaluating environmental sensitivity at the basin scale through the use of geographic information systems and remotely sensed data: an example covering the Agri basin - Southern Italy. Catena, 40, 19-35.

Berling-Wolf, S. and Wu, J. (2004). Modelling urban landscape dynamics: a case study in Phoenix USA. Urban Ecosystems, 7, 215-240.

Bruegmann, R. (2005). *Sprawl: A Compact History*. University of Chicago Press, Chicago.

Burchell, R., Downs, A., McCann, B. and Mukherji, S. (2005). *Sprawl Costs: Economic Impacts of Unchecked Development*. Island Press, Washington, DC.

Cakir, G., Un, C., Baskent, E.Z., Kose, S., Sivrikaya, F. and Keles, S. (2008). Evaluating urbanization, fragmentation and land use/cover change pattern in Istanbul city, Turkey from 1971 to 2002. Land Degradation & Development, 19, 663-675.

Carone, M.T., Simoniello, T., Manfreda, S. and Caricato, G. (2009). Watershed influence on fluvial ecosystems: an integrated methodology for river water quality management. Environmental Monitoring and Assessment, 152(1), 327-342.

Catalàn, B., Saurì, D. and Serra, P. (2008). Urban sprawl in the Mediterranean? Patterns of growth and change in the Barcelona Metropolitan Region 1993–2000. Landscape and Urban Planning, 85(3-4), 174-184.

Chorianopoulos, I., Pagonis, T., Koukoulas, S. and Drymoniti, S. (2010). Planning, competitiveness and sprawl in the Mediterranean city: the case of Athens. Cities, 27, 249-259.

Colantoni, A., Mavrakis, A., Sorgi, T. and Salvati, L. (2015). Towards a 'polycentric' landscape? Reconnecting fragments into an integrated network of coastal forests in Rome. Rendiconti Lincei, 26(3), 615-624.

Colantoni, A., Delfanti, L., Recanatesi, F., Tolli, M. and Lord, R. (2016). Land use planning for utilizing biomass residues in Tuscia Romana (central Italy): preliminary results of a multi criteria analysis to create an agro-energy district. Land Use Policy, 50, 125-133.

Coluzzi, R., D'Emilio, M., Imbrenda, V., Giorgio, G.A., Lanfredi, M., Macchiato, M., Simoniello, T. and Telesca, V. (2019). Investigating climate variability and long-term vegetation activity across heterogeneous Basilicata agroecosystems. Geomatics, Natural Hazards and Risk, 10(1), 168-180.

Conacher, A.J. and Sala, M. (1998). *Land Degradation in Mediterranean Environments of the World*. Wiley, Chichester.

Couch, C., Petschel-held, G. and Leontidou, L. (2007). *Urban Sprawl in Europe: Landscapes, Land-Use Change and Policy*. Blackwell, London.

D'Emilio, M., Coluzzi, R., Macchiato, M., Imbrenda, V., Ragosta, M., Sabia, S. and Simoniello, T. (2018). Satellite data and soil magnetic susceptibility measurements for heavy metals monitoring: findings from Agri Valley (Southern Italy). Environmental Earth Sciences, 77(3), 1-7.

De Montis, A., Martín, B., Ortega, E., Ledda, A. and Serra, V. (2017). Landscape fragmentation in Mediterranean Europe: a comparative approach. Land Use Policy, 64, 83-94.

De Rosa, M., McElwee, G. and Smith, R. (2019). Farm diversification strategies in response to rural policy: a case from rural Italy. Land Use Policy, 81, 291-301.

DiBari, J. (2007). Evaluation of five landscape-level metrics for measuring the effects of urbanization on landscape structure: the case of Tucson, Arizona, USA. Landscape and Urban Planning, 79, 308-313.

Di Felice, V., Bessa, E.R.D.J.S., Mancinelli, R., Batista, J.G.F. and Campiglia E. (2014). Rurality and agroecosystem sustainability: a case study at farm-field level in Terceira Island (Portugal) and in Viterbo Province (Italy). Renewable Agriculture and Food Systems, 29(3), 265-276.

Di Ludovico, D. and D'Ascanio, F. (2019). European cross-scale spatial planning and territorial frames in the Italian Median Macroregion. European Planning Studies, 27(7), 1369-1390.

Ferrara, A., Salvati, L., Sabbi, A. and Colantoni, A. (2014). Urbanization, soil quality and rural areas: towards a spatial mismatch? Science of the Total Environment, 478, 116-22.

Fiorini, L., Zullo, F., Marucci, A. and Romano, B. (2019). Land take and landscape loss: Effect of uncontrolled urbanization in Southern Italy. Journal of Urban Management, 8(1), 42-56.

Fregolent, L. and Vettoretto, L. (2017). Land use regulation and social regulation: an unexplored link. Some reflections on the origins and evolution of sprawl in the Veneto "città diffusa". Land Use Policy, 69, 149-159.

García-Ruiz, J.M., Lasanta, T., Nadal-Romero, E., Lana-Renault, N. and Álvarez-Farizo, B. (2020). Rewilding and restoring cultural landscapes in Mediterranean mountains: opportunities and challenges. Land Use Policy, 99, 104850.

Gargiulo, Morelli, V. and Salvati, L. (2010). *Ad Hoc Urban Sprawl in the Mediterranean City: Dispersing a Compact Tradition?* Nuova Cultura, Rome.

Gariano, S.L., Petrucci, O., Rianna, G., Santini, M. and Guzzetti, F. (2018). Impacts of past and future land changes on landslides in southern Italy. Regional Environmental Change, 18(2), 437-449.

Gobattoni, F., Pelorosso, R., Leone, A. and Ripa, M.N. (2015). Sustainable rural development: The role of traditional activities in Central Italy. Land Use Policy, 48, 412-427.

Goliński, P., Czerwiński, M., Jørgensen, M., Mølmann, J.A., Golińska, B. and Taff, G. (2018). Relationship between climate trends and grassland yield across contrasting European locations. Open Life Sciences, 13(1), 589-598.

Gullino, P., Battisti, L. and Larcher, F. (2018). Linking multifunctionality and sustainability for valuing peri-urban farming: a case study in the Turin Metropolitan Area (Italy). Sustainability, 10(5), 1625.

Gupta, R. and Mishra, A. (2019). Climate change induced impact and uncertainty of rice yield of agro-ecological zones of India. Agricultural Systems, 173, 1-11.

Imbrenda, V., D'Emilio, M., Lanfredi, M., Ragosta, M. and Simoniello, T. (2013). Indicators of land degradation vulnerability due to anthropic factors: tools for an efficient planning. In *Sustainable Practices: Concepts, Methodologies, Tools, and Applications* (pp. 1400-1413). IGI Global, Pennsylvania.

Imbrenda, V., D'Emilio, M., Lanfredi, M., Macchiato, M., Ragosta, M. and Simoniello, T. (2014). Indicators for the estimation of vulnerability to land degradation derived from soil compaction and vegetation cover. European Journal of Soil Science, 65(6), 907-923.

Imbrenda, V., Coluzzi, R., Lanfredi, M., Loperte, A., Satriani, A. and Simoniello, T. (2018). Analysis of landscape evolution in a vulnerable coastal area under natural and human pressure. Geomatics, Natural Hazards and Risk, 9(1), 1249-1279.

Ioannidis, C., Psaltis, C. and Potsiou, C. (2009). Towards a strategy for control of suburban informal buildings through automatic change detection. Computers, Environment and Urban Systems, 33, 64-74.

Iosifides T. and Politidis T. (2005). Socio-economic dynamics, local development and desertification in western Lesvos, Greece. Local Environment, 10, 487-499.

ISTAT (2011). *15° Censimento della Popolazione e della Abitazione*. Istituto Nazionale di Statistica, Rome.

Jun, H. and Xiang, H. (2011). Development of circular economy is a fundamental way to achieve agriculture sustainable development in China. Energy Procedia, 5, 1530-1534.

Kang, Y., Khan, S. and Ma, X. (2009). Climate change impacts on crop yield, crop water productivity and food security–A review. Progress in Natural Science, 19(12), 1665-1674.

Knickel, K., Redman, M., Darnhofer, I., Ashkenazy, A., Chebach, T.C., Šūmane, S. *et al.* (2018). Between aspirations and reality: making farming, food systems and rural areas more resilient, sustainable and equitable. Journal of Rural Studies, 59, 197-210.

Kreiling, R.M., Thoms, M.C. and Richardson, W.B. (2018). Beyond the edge: linking agricultural landscapes, stream networks, and best management practices. Journal of Environmental Quality, 47(1), 42-53.

Lagacherie, P., Álvaro-Fuentes, J., Annabi, M., Bernoux, M., Bouarfa, S., Douaoui, A. *et al.* (2018). Managing Mediterranean soil resources under global change: expected trends and mitigation strategies. Regional Environmental Change, 18(3), 663-675.

La Rosa, D., Barbarossa, L., Privitera, R. and Martinico, F. (2014). Agriculture and the city: a method for sustainable planning of new forms of agriculture in urban contexts. Land Use Policy, 41, 290-303.

Le Bivic, C. and Melot, R. (2020). Scheduling urbanization in rural municipalities: local practices in land-use planning on the fringes of the Paris region. Land Use Policy, 99, 105040.

Lopez-Goyburu, P. and García-Montero, L.G. (2018). The urban-rural interface as an area with characteristics of its own in urban planning: a review. Sustainable Cities and Society, 43, 157-165.

Luck, M. and Wu, J. (2002). A gradient analysis of urban landscape pattern: a case study from the Phoenix metropolitan region, Arizona, USA. Landscape Ecology, 17, 327-339.

Malek, Ž., Verburg, P.H., Geijzendorffer, I.R., Bondeau, A. and Cramer, W. (2018). Global change effects on land management in the Mediterranean region. Global Environmental Change, 50, 238-254.

Marraccini, E., Debolini, M., Moulery, M., Abrantes, P., Bouchier, A., Chéry, J.P. *et al.* (2015). Common features and different trajectories of land cover changes in six Western Mediterranean urban regions. Applied Geography, 62, 347-356.

Meeus, J.H.A. (1993). The transformation of agricultural landscapes in Western Europe. The Science of the Total Environment, 129, 171-190.

Mueller, L., Eulenstein, F., Dronin, N.M., Mirschel, W., McKenzie, B.M., Antrop, M.,... and Poulton, P. (2021). Agricultural landscapes: history, status and challenges. In L. Mueller, V.G. Sychev, N.M. Dronin and F. Eulenstein [eds.]. *Exploring and Optimizing Agricultural Landscapes* (pp. 3-54). Springer, Cham.

Munafò, M., Assennato, F., Congedo, L., Luti, T., Marinosci, I., Monti, G., Riitano, N., Sallustio, L., Strollo, A., Tombolini, I. and Marchetti, M. (2015). *Il Consumo di Suolo in Italia Rapporti 218/2015*. ISPRA, Roma.

Muscio, A. and Sisto, R. (2020). Are agri-food systems really switching to a circular economy model? Implications for European research and innovation policy. Sustainability, 12(14), 5554.

Næss, P., Saglie, I.L. and Richardson, T. (2020). Urban sustainability: is densification sufficient?. European Planning Studies, 28(1), 146-165.

Nickayin, S.S., Salvati, L., Coluzzi, R., Lanfredi, M., Halbac-Cotoara-Zamfir, R., Salvia, R., Quaranta G., Alhuseen, A. and Gaburova, L. (2021). What happens in the city when long-term urban expansion and (un) sustainable fringe development occur: the case study of Rome. ISPRS International Journal of Geo-Information, 10(4), 231.

Nolè A., Law B.E., Magnani F., Matteucci G., Ferrara A., Ripullone, F. *et al.* (2009). Application of the 3-PGS model to assess carbon accumulation in forest ecosystems at a regional level. Canadian Journal of Forest Research, 39(9), 1647-1661.

Pa, S.-Y., Du M.A., Huang I.-T., Liu I.-H., Chang E.-E. and Chiang, P.-C. (2015). Strategies on implementation of waste-to-energy (WTE) supply chain for circular economy system: a review. Journal of Cleaner Production, 108, 409-421.

Palazzo, A.L. and Aristone, O. (2017). Peri-urban matters. Changing olive growing patterns in central Italy. Sustainability, 9(4), 638.

Paul, V. and Tonts, M. (2005). Containing urban sprawl: trends in land use and spatial planning in the Metropolitan Region of Barcelona. Journal of Environmental Planning and Management, 48(1), 7-35.

Polidoro, M., De Lollo, J.A. and Barros, M.V.F. (2012). Urban sprawl and the challenges for urban planning. Journal of Environmental Protection, 3(09), 1010-1019.

Polyzos, S., Christopoulou, O., Minetos, D. and Leal Filho, W. (2008). An overview of urban–rural land use interactions in Greece. International Journal of Agricultural Resources, Governance and Ecology, 7, 276-296.

Poponi, S., Arcese, G., Mosconi, E.M., Pacchera, F., Martucci, O. and Elmo, G.C. (2021). Multi-actor governance for a circular economy in the agri-food sector: bio-districts. Sustainability, 13(9), 4718.

Quaranta, G., Salvia, R., Salvati, L., Paola, V.D., Coluzzi, R., Imbrenda, V. and Simoniello, T. (2020). Long term impacts of grazing management on land degradation in a rural community of Southern Italy: depopulation matters. Land Degradation & Development, 31(16), 2379-2394.

Romano, B., Fiorini, L., Marucci, A. and Zullo, F. (2020). The urbanization run-up in Italy: from a qualitative goal in the boom decades to the present and future unsustainability. Land, 9(9), 301.

Rubio, J.L. and Bochet, E. (1998). Desertification indicators as diagnosis criteria for desertification risk assessment in Europe. Journal of Arid Environments, 39, 113-120.

Salvati, L. (2014a). A socioeconomic profile of vulnerable land to desertification in Italy. Science of the Total Environment, 466-467, 287-299.

Salvati, L. (2014b). Agro-forest landscape and the 'fringe' city: a multivariate assessment of land-use changes in a sprawling region and implications for planning. Science of the Total Environment, 490, 715-723.

Salvati, L. and Carlucci, M. (2011). The economic and environmental performances of rural districts in Italy: are competitiveness and sustainability compatible targets? Ecological Economics, 70(12), 2446-2453.

Salvati, L., Bajocco, S., Ceccarelli, T., Zitti, M. and Perini, L. (2011). Towards a process-based evaluation of land vulnerability to soil degradation in Italy. Ecological Indicators, 11, 1216-1227.

Salvati, L. and Carlucci, M. (2016). Patterns of sprawl: the socioeconomic and territorial profile of dispersed urban areas in Italy. Regional Studies, 50(8), 1346-1359.

Salvati, L., Ranalli, F. and Gitas, I. (2014). Latent exurban development: city expansion along the rural-to-urban gradient in growing and declining regions of Southern Europe. Landscape fragmentation and the agro-forest ecosystem along the urban gradient: an exploratory study. International Journal of Sustainable Development & World Ecology, 21(2), 160-167.

Salvati, L. and Sabbi A. (2011). Exploring long-term land cover changes in an urban region of southern Europe. International Journal of Sustainable Development & World Ecology, 18(4), 273-282.

Salvati, L., Serra, P., Rugiero, S. and Sabbi, A. (2016). Complexity in action? Fringe agro-forest systems, demography and societal transformations in Mediterranean peri-urban areas. Agriculture and Agricultural Science Procedia, 8, 222-227.

Salvati, L. and Zitti, M. (2009). Multivariate analysis of socio-economic indicators to estimate land degradation sensitivity: a case study applied to a Mediterranean area. International Journal of Ecological Economics and Statistics, 15(F09), 93-102.

Satriani, A., Loperte, A., Imbrenda, V. and Lapenna, V. (2012). Geoelectrical surveys for characterization of the coastal saltwater intrusion in Metapontum forest reserve (Southern Italy). International Journal of Geophysics, 2012.

Scherer, L.A., Verburg, P.H. and Schulp, C.J. (2018). Opportunities for sustainable intensification in European agriculture. Global Environmental Change, 48, 43-55.

Secondi, L., Principato, L., Ruini, L. and Guidi, M. (2019). Reusing food waste in food manufacturing companies: the case of the tomato-sauce supply Chain. Sustainability, 11(7), 2154.

Serra, P., Vera, A. and Tulla A.F. (2014). Spatial and socio-environmental dynamics of Catalan regional planning from a multivariate statistical analysis using 1980s and 2000s data. European Planning Studies, 22(6), 1280-1300.

Seto, K.C. and Fragkias, M. (2005). Quantifying spatiotemporal patterns of urban land-use change in four cities of China with time series landscape metrics. Landscape Ecology, 20, 871-888.

Shamsuzzoha, A.H.M., Grant, A. and Clarke, J. (2012). Implementation of renewable energy in Scottish rural area: a social study. Renewable & Sustainable Energy Reviews, 16, 185-191.

Shaw, B.J., van Vliet, J. and Verburg, P.H. (2020). The peri-urbanization of Europe: a systematic review of a multifaceted process. Landscape and Urban Planning, 196, 103733.

Silvestri, N. and Bonari, E. (2010). Agricultural management in peri-urban areas. The experience of an international workshop, in Felici Editore, Pisa, Sustainability of peri-urban agriculture. Peri-urban agriculture and water quality. Phosphorus pollution in Lake Massaciuccoli, pp. 141-150.

Smetanová, A., Follain, S., David, M., Ciampalini, R., Raclot, D., Crabit, A. and Le Bissonnais, Y. (2019). Landscaping compromises for land degradation neutrality: the case of soil erosion in a Mediterranean agricultural landscape. Journal of Environmental Management, 235, 282-292.

Solly, A., Berisha, E. and Cotella, G. (2021). Towards sustainable urbanization. learning from what's out there. Land, 10(4), 356.

Song, Y. and Knaap, G.J. (2004). Measuring urban form. Is Portland winning the war on sprawl? Journal of the American Planning Association, 70, 210-225.

Spyra, M., La Rosa, D., Zasada, I., Sylla, M. and Shkaruba, A. (2020). Governance of ecosystem services trade-offs in peri-urban landscapes. Land Use Policy, 95, 104617.

Statuto, D., Frederiksen, P. and Picuno, P. (2019). Valorization of agricultural by-products within the "Energyscapes": renewable energy as driving force in modeling rural landscape. Natural Resources Research, 28(1), 111-124.

Tziolas, E., Manos, B. and Bournaris, T. (2016). Planning of agro-energy districts for optimum farm income and biomass energy from crops residues. Operational Research - An International Journal (ORIJ), pp. 1-12.

van Vliet, J., Birch-Thomsen, T., Gallardo, M., Hemerijckx, L.M., Hersperger, A.M., Li, M., *et al.* (2020). Bridging the rural-urban dichotomy in land use science. Journal of Land Use Science, 15(5), 585-591.

Varela-Candamio, L., Morollón, F.R. and Sedrakyan, G. (2019). Urban sprawl and local fiscal burden: analysing the Spanish case. Empirica, 46(1), 177-203.

Varga, D., Vila Subirós, J., Barriocanal, C. and Pujantell, J. (2018). Landscape transformation under global environmental change in Mediterranean mountains: agrarian lands as a guarantee for maintaining their multifunctionality. Forests, 9(1), 27.

Weber C., Petropoulou C. and Hirsch J. (2005). Urban development in the Athens metropolitan area using remote sensing data with supervised analysis and GIS. International Journal of Remote Sensing, 26(4), 785-796.

Wilson, G.A. and Juntti, M. (2005). *Unraveiling Desertification. Policies and Actor Networks in Southern Europe*. Wageningen Academic Publishers, Wageningen.

Zabaniotou, A., Rova, D., Libutti, A. and Monteleone, M. (2015). Boosting circular economy and closing the loop in agriculture: case study of a small-scale pyrolysis–biochar based system integrated in an olive farm in symbiosis with an olive mill. Environmental Development, 14, 22-36.

Zimmerer, K.S., Jiménez-Olivencia, Y., Ruiz-Ruiz, A. and Porcel-Rodríguez, L. (2020). Agri-food land transformations and immigrant farm workers in peri-urban areas of Spain and the Mediterranean. Land, 9(12), 472.

# 6

# Like Father Like Son? An Analysis of Opportunity and Discrimination in Tertiary Education for OECD Countries

**Mariateresa Ciommi, Francesca Mariani, Maria Cristina Recchioni, Gloria Polinesi, and Giuseppe Ricciardo Lamonica**

Department of Economic and Social Sciences, Marche Polytechnic University, Ancona, Italy

## Abstract

The existence of a relationship between education level and the socio-economic results has been demonstrated and investigated in depth. It is well-known that a higher level of study is associated with better levels of health, a higher probability of obtaining a job, and higher perceived incomes. For these reasons, interest in analysing academic performance has grown in recent years. Data from the *Education at a Glance* annual report shows that there is great disparity in academic performance in OECD countries, particularly in terms of tertiary education. It emerges that more adults between the ages of 25 and 34 are attaining tertiary degrees than adults between 55 and 64. Moreover, for individuals aged 30–34 years who achieve a tertiary level of education, adults with at least one tertiary-educated parent are still more likely to attain a tertiary degree than individuals from families whose parents do not have a tertiary level of education.

To measure this phenomenon, two aspects must be analysed: mobility and opportunity. "Opportunity" means the probability for a child to attain tertiary education regardless of the level of education achieved by the parents. "Mobility" implies the similarity in probabilities of achieving tertiary education between children from families with a different level of education. We introduce an index for each of these aspects and we rank the OECD countries according to these two indices. A comparison of the two rankings reveals conflicting results in some cases. Countries that occupy high positions in one ranking do not necessarily occupy equally high positions in the other. To overcome this contradiction, we propose seven different aggregate methods to combine the two indices in order to provide a ranking that accounts for both dimensions.

## 6.1 Introduction

Education achievement and socio-economic outcomes are closely related. Higher levels of educational achievement are associated with improved health, higher salaries and incomes, and greater possibilities for establishment in the labour market. However, levels of educational achievement are handed down among generations. Children living in families with university-educated parents obtain the same title more often than those living in families with parents that have lower educational qualifications (see [10], among others). In particular, the OECD survey highlights how parents' educational attainment is a much stronger predictor of an individual's education attainment than age or gender. Moreover, educational attainment is significantly correlated across generations. This implies that educational traits persist between generations in all OECD countries. In contrast, the share of individuals with at least one university-educated parent varies significantly across countries. For instance, Italy and Turkey show a lower percentage (5% and 4%, respectively), while it is higher in Israel (43%), Canada, and New Zealand (42%) (see OECD, 2017).

Universities, therefore, play an important role in ensuring real social mobility. As a result, government policies should support universities to promote social mobility by encouraging student recruitment, especially students belonging to disadvantaged backgrounds.

Nevertheless, low upward mobility does not necessarily indicate a lower opportunity to attain high levels of education. For example, upward mobility may be low in countries where a large share of parents has already attained tertiary education.

To evaluate how a family's educational level matters for a given individual (child) in obtaining the maximum educational title, we analyse 26 countries, namely, Australia, Austria, Canada, Chile, Denmark, Estonia, Finland, France, Germany, Japan, Greece, Ireland, Israel, Italy, Korea, Norway, New Zealand, Holland, Poland, Czech Republic, Slovakia, Slovenia, Spain, United States, Sweden, and Turkey. We use OECD data (OECD, 2017) from 2015, focusing on aggregate (country level) data for individuals between the ages of 30 and 44.

We introduce two indices: mobility and opportunity. The first index is defined as the difference in probability of achieving tertiary education between children from families characterized by different levels of education. Opportunity refers to the probability for a child to achieve tertiary education regardless of the level of education achieved by the parents.

We, therefore, rank the countries separately according to the two dimensions. Of course, the two rankings do not necessary coincide; they capture two different aspects like two sides of the same coin. Hence, we develop a new index that addresses the two dimensions simultaneously.

There is a large amount of literature on the construction of composite indicators related to social, economic, or environmental domains. In recent years, there has been increasing interest in collapsing all the information into a single number mainly because it allows the performance of a country to be measured and compared over time and space in a very fast and intuitive way. Moreover, policy makers, a single number helps to support political decisions and favour public communication (OECD, 2008). While it is true that collapsing all the information into a single number might hide relevant aspects (Bleys, 2012), there appear to be more advantages in doing so than disadvantages.

Thus, to achieve our aim, we propose seven different aggregate methods: the simple arithmetic mean (EW*)*, the adjusted Mazziotta–Pareto index (AMPI), and modifications thereof based on weighting systems depending on the Gini coefficient, its reciprocal (GW, iG, GAMPI, and iGAMPI), and the geometric mean (GM).[1] A comparison of the corresponding rankings highlights the countries with the best and worst performances. Furthermore, by analysing the rankings, we identify the method with the highest correlations, which, therefore, represents all the other methods. This allows us to suggest to policy makers the best method to aggregate our data. Finally, we compare the ranking obtained using this method with expenditure on educational institutions as a percentage of gross domestic product (GDP) per level of education to see if there is a connection between input (expenditure) and output (opportunity and mobility in education).

The remainder of the paper is organized as follows. In Section 6.2, we present OECD data and introduce the basic notations and definitions. In Section 6.3, we compute the indices of mobility and opportunity. We review the indices proposed in the literature and discuss the desired properties for a mobility index. Section 6.4 illustrates the seven aggregation methods

[1] See Ciommi *et al.* (2017) for a detailed description of the first six methods.

according to mobility and opportunity. The corresponding rankings, a comparison among methods, and the relationship with expenditure are reported in Section 6.5. Finally, Section 6.6 presents the conclusions.

## 6.2 Data and Basic Notation

### 6.2.1 Data

We use country-level data for 26 OECD countries, namely Australia, Austria, Canada, Chile, Czech Republic, Denmark, Estonia, Finland, France, Germany, Greece, Ireland, Israel, Italy, Japan, Korea, Netherlands, New Zealand, Norway, Poland, Slovak Republic, Slovenia, Spain, Sweden, Turkey, and the United States. The data is from the Survey of Adult Skills under the OECD Programme for the International Assessment of Adult Competencies (PIAAC) (OECD, 2017).

Among the various data collected, we focus on individuals who are 30–44 years old. We concentrate our attention on the following percentages: (i) adults who have attained tertiary education and whose parents both have less than tertiary education; (ii) adults who have attained tertiary education and who have at least one parent who attained tertiary education; (iii) adults whose parents both have less than tertiary education; and (iv) adults who have at least one parent who attained tertiary education.

Moreover, we also consider data on expenditure on educational institutions as a percentage of GDP for different levels of education with a focus on tertiary education.

### 6.2.2 Basic definitions and notation

We summarize the information on the education achievements of parents and children through a 2 × 2 matrix, denoted by $E$. The rows of the matrix indicate the education of the parents, whereas the columns represent the children. In particular, $PL$ denotes the event "to have at least one parent tertiary educated"; $\overline{PL}$, the complement of $PL$, represents the event "to have neither parent tertiary educated"; $CL$ denotes the event "to have achieved tertiary education" and $\overline{CL}$ indicates the complement of $CL$, the event "to have not achieved tertiary education". Consequently, the columns display the percentage of children who have achieved tertiary education ($CL$) or not ($\overline{CL}$). Likewise, the rows indicate the percentage of parents (at least one) who have achieved tertiary education ($PL$) or not ($\overline{PL}$).

We use $p$ to denote the percentage of children who have not achieved a tertiary education level given that their parents have not achieved the tertiary education level, $p = P(\overline{CL}|\overline{PL})$.

In the same way, $p$ represents the probability that a child has not achieved a tertiary education level given that he/she has at least one parent who has achieved a tertiary education level, $q = P(\overline{CL}|PL)$.

Formally:

$$E = \begin{pmatrix} P(\overline{CL}|\overline{PL}) & 1 - P(\overline{CL}|\overline{PL}) \\ P(\overline{CL}|PL) & 1 - P(\overline{CL}|PL) \end{pmatrix} = \begin{pmatrix} p & 1-p \\ q & 1-q \end{pmatrix}.$$

This can be interpreted as a sort of transition matrix and the rows sum to one.

In fact, we observe that this is a particular case of a $n \times n$ transition matrix with $n = 2$. Thus, starting from the definition of intergenerational mobility as reported in Van de Gaer *et al.* (2001), we assume that we have two groups, parents and children, divided according to their level of education, which more specifically consists of only two states: "with tertiary education" and "without tertiary education". Again, the parents' status is in the rows and the columns denote the children's status. Thus, the matrix can be interpreted as an intergenerational transition matrix since we are analysing the educational status of parents (time $t - 1$) and children (time $t$). We stress that, in general, intergenerational mobility refers to income levels, where the income distribution of both parents and children is usually divided into $n$ quantiles.

Moreover, here, we do not require the matrix $E$ be bistocastic, that is, both rows and columns sum to one; rather, we are working with row-stochastic matrices.

The following example clarifies our definition. We construct the matrix associated with Italy:

$$E_{\text{ITA}} = \begin{pmatrix} e_{1,1} & e_{1,2} \\ e_{2,1} & e_{2,2} \end{pmatrix} = \begin{pmatrix} 86.14\% & 13.86\% \\ 32.32\% & 67.68\% \end{pmatrix}.$$

The quantity 86.14% represents the percentage of adults who have not achieved tertiary education from families where neither parent has achieved tertiary education. This means that the remaining 13.86% has achieved tertiary education. Likewise, the bottom row represents the percentage of adults from families where at least one parent has achieved tertiary education, where

32.32% of children have not achieved tertiary education, and the remaining 67.68% have. It is easy to see that the numbers in each row sum to one. Thus, $E$ is a row-stochastic matrix.

Before moving on to further definitions and computations, we first look at the data for Italy. What is surprising is the very high value of $p$. Looking at the OECD data, we find that it ranges from 49.73% to 87.85%. Comparing data for different countries, we see that Italy displays the second worst situation. Italy follows Turkey with 87.85% and precedes the Czech Republic with 84.80%. This means that in countries like Turkey, Italy, and the Czech Republic, it is quite hard for children to achieve tertiary education if their parents lack the title. Canada and Korea display the lowest percentages, 49.73% and 51.92%, respectively. This means that in these countries, the parents' education level does not significantly affect children's achievement of the title and, implicitly, that there is higher educational mobility. Looking at the values of $q$, which, as mentioned before, represents the probability that a child does not achieve a tertiary education level given that at least one parent who achieved a tertiary education level, we find that the values range

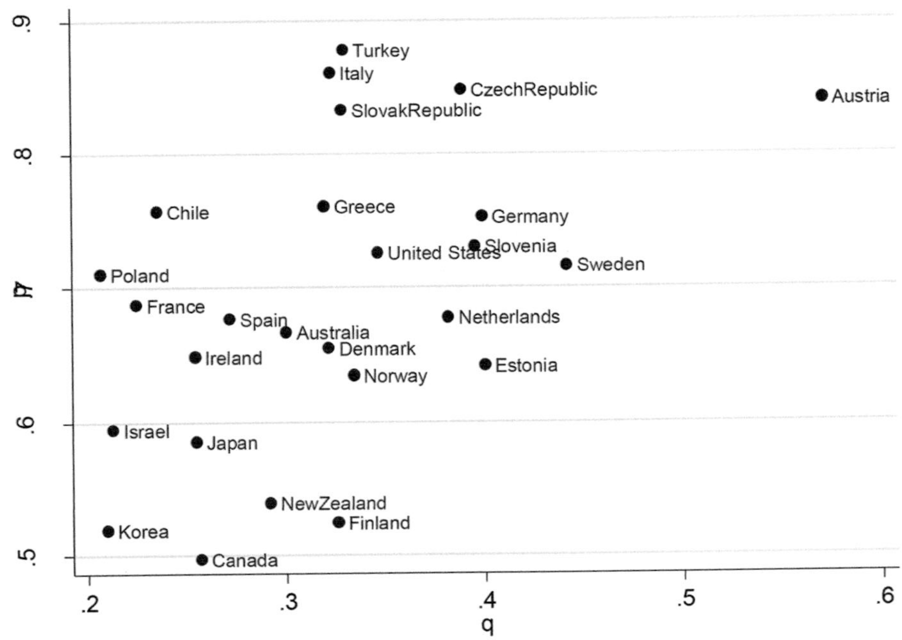


**Figure 6.1** p vs q.

from 20.70% to 57.10%. Half of countries are concentrated in the interval (30%–40%) and Poland, Korea, and Israel have the lowest values – 20.70%, 21.00%, and 21.30%, respectively. This means that only about 20% of children whose parents have tertiary education do not achieve tertiary education; the remaining 80% graduate. At the bottom of the ranking, we have countries such as Austria (57.10%), Sweden (44.18%), and Estonia (40.03%).

Figure 6.1 shows a comparison of $p$ and $q$ values. The correlation between these two aspects is not very high ($\rho = 0.4671$).

The percentages of individuals whose educational status does not change with respect to their parents, namely "stayers", are on the diagonal, while those off the diagonal are called "movers".

Without a loss of generality, we assume $p > q$. This assumption is also confirmed by our data, since the mean and ranges of $p$ and $q$ are, respectively, 0.6907731 and 0.32035 and [0.4972,0.8784] and [0.2070,0.5710], meaning that the majorization holds.

## 6.3 Mobility and Opportunity: Two Sides of the Same Coin

### 6.3.1 Discrimination Index

We define the discrimination index ($\Delta$) associated with the matrix as the difference in probability of achieving tertiary education for children from families with different levels of education. Formally, we have

$$\Delta = \Pr(CL|PL) - \Pr(CL|\overline{PL}) = (1 - q) - (1 - p) = p - q.$$

Looking at the empirical matrices, we have: $p > q$; so $\Delta$ is always strictly positive. A value of $\Delta$ approaching 0 means that parents' education does not determine a child's educational achievement. In other words, the parent's educational level has no effect on the child's educational level, corresponding to a situation of perfect mobility. When the index reaches the (theoretical) value of zero, $p = q$ and we have null discrimination among the children coming from families with a low or high educational level. For this reason, we refer to it as the best case.

On the other hand, positive values of the discrimination index, $\Delta > 0$, describe societies in which children are more likely to graduate if at least one parent is a university graduate. From a mathematical point of view, the highest discrimination is achieved when the transition matrix coincides with the identity matrix, that is, $p=1-q=1$. In this case, we have no intergenerational transition between educational levels, and we refer to it as the worst case.

#### 6.3.1.1 Mobility and discrimination

The literature on mobility has introduced several measures based on the transition matrix. The construction of such indices enables cross-national comparison since countries can be ranked according to the computed indices. Over the years, many indices have been proposed. For instance, assuming that there are *n* different classes (or states) in a given country, the transition matrix *P* is an $n \times n$ matrix and $p_{ij}, i,j=1,\ldots,n$ denotes a generic element representing the probability that a child with a parent whose income falls in quantile *i* ends up with an income in quantile *j* (Van de Gaer *et al.*, 2001).

Shorrocks (1978) defined two indices capturing mobility: the *trace index (T)* and the *determinant measure (D)*, namely:

$$T = \frac{n - \mathrm{Tr}(P)}{n-1} \qquad D = 1 - \frac{\det(P)}{n-1},$$

where Tr(*P*) and det(*P*) denote, respectively, the trace and the determinant of transition matrix *P*.

The trace index reflects the concentration of the transition matrix around the diagonal. In fact, as pointed out by Shorrocks, in a transition matrix, the highest values seem to be clustered around the main diagonal.

Similarly, Conlisk and Sommers (1979) introduce a mobility index based on the eigenvalues of the transition matrix, defined as,

$$\Lambda = 1 - \lambda,$$

where $\lambda$denotes the second largest eigenvalue of *P*, while the first is equal to one by virtue of the properties of the transition matrix.

We observe that for the special case of $n = 2$, the transition matrix *P* reduces to the matrix *E* defined above. Thus, it is easy to prove that in this particular situation, where there are only two states, the mobility indices *T*, *D*, and $\Lambda$reduce to the same values:

$$T = D = \Lambda = 1 - (p - q).$$

Moreover, in this specific case, the above-mentioned indices are closely related to the discrimination index via the following equation:

$$T = D = \Lambda = 1 - (p - q).$$

The discrimination index $\Delta$can, therefore, be interpreted as an "immobility index". In other words, the lowest discrimination, which in turn corresponds to the highest mobility, is achieved if all rows of the transition matrix are identical. In this way, discrimination can be interpreted as the *opposite* of the concept of mobility. Consequently, since we expect that the greater the value, the better the situation, we compute the complement to 1 of the discrimination index. We stress that this is just a transformation, the sole effect of which is to overturn the ordering induced by the index. With this transformation, higher values correspond to better situations. Thus, the greater the value of $\Delta$ the lower the social mobility. We denote the new index with

$$\mathrm{MI} = 1 - \Delta.$$

#### 6.3.1.2 Properties of the index

Following Shorrocks (1978), we introduce properties that might be satisfied by an index that simultaneously accounts for discrimination and mobility.

We remember that the index should be a continuous real function, denoted by $I$, defined over the set of (transition) matrices $E$. We restrict the range of the index to the interval [0, 1] to favour comparisons. Thus, we introduce a *normalization* axiom:

**Axiom 1** [*Normalization*] (*N*): $0 \le I(E) \le 1$.

This is a common requirement for indices. It does not impose significant constraints on the set of potential measures but rather rescales the indices.

According to the classical measure of mobility, the probability of movement between classes is given by the off-diagonal elements of the transition matrix. Here, we have $1 - p$ and $q$ which represent the movements between classes, that is, the transition from a parent without tertiary education to a child with tertiary education (element $e_{12}$) and from a parent with tertiary education to a child without tertiary education (element $e_{21}$).

A second natural requirement is that the index should increase if there is an increase in the off-diagonal elements at the expense of the diagonal elements, which indicates immobility, namely the fact that people remain in the same status between generations. Thus, we compare two situations, denoted by means of two matrices, $E$ and $E'$. We affirm that $E \geq E'$ if $p_{ij} \geq p'_{ij}$ for all $i \neq j$ and $p_{ij} > p'_{ij}$ for some/all $i \neq j$. Formally,

**Axiom 2** [*Monotonicity*] (*M*): $E \geq E'$ implies that $I(E) \geq I(E')$.

Moreover, we define an index that accounts for two extreme situations, that is, the index assumes a value of zero in the worse situation and one in the best case.

**Axiom 3** [*Worst case*] (*W*): *I(E)* = 0

**Axiom 4** [*Best case*] (*B*): *I(E)* = 1

Axiom 3 implies that has identical rows while Axiom 4 ensures that the index is one when matrix coincides with the identity matrix.

It is easy to prove that the index of mobility we have introduced satisfies the above-mentioned properties.

#### 6.3.1.3 Ranking OECD countries by mobility

We computed the *MI* for 26 OECD countries. The index ranges from 0.4514 to 0.8012, with an average value of 0.6296 and a standard deviation of 0.1016. Moreover, there are 16 countries with values greater than the average, which justifies the fact that the data display a negative coefficient of asymmetry (–0.25).

Table 6.1 shows the mobility index and corresponding rank. Countries are ranked from most to least "mobile", that is, countries in the first positions exhibit the best performance.

Finland has the highest level of mobility, meaning that parents' education does not (negatively) affect the probability for a child to reach tertiary education. Children present almost the same probability to graduate whether their parents have or do not have tertiary education.

**Table 6.1** Ranking according to mobility.

| Rank | Country | Mobility | Rank | Country | Mobility |
|---|---|---|---|---|---|
| 1 | Finland | 0.8012 | 14 | Australia | 0.6337 |
| 2 | Canada | 0.7596 | 15 | United States | 0.6211 |
| 3 | Estonia | 0.7591 | 16 | Israel | 0.6184 |
| 4 | New Zealand | 0.7523 | 17 | Ireland | 0.6056 |
| 5 | Austria | 0.7309 | 18 | Spain | 0.5956 |
| 6 | Sweden | 0.7262 | 19 | Greece | 0.5589 |
| 7 | Netherlands | 0.7047 | 20 | Czech Republic | 0.5410 |
| 8 | Norway | 0.6996 | 21 | France | 0.5379 |
| 9 | Korea | 0.6908 | 22 | Poland | 0.4967 |
| 10 | Japan | 0.6691 | 23 | Slovak Republic | 0.4959 |
| 11 | Denmark | 0.6670 | 24 | Chile | 0.4785 |
| 12 | Slovenia | 0.6653 | 25 | Italy | 0.4618 |
| 13 | Germany | 0.6467 | 26 | Turkey | 0.4514 |

Canada and Estonia occupy the second and the third positions, with very similar values. At the bottom of the ranking, we find Chile, Italy, and Turkey with 0.4785, 0.4618, and 0.4514, respectively. For these countries, the parents' education has a great impact on the probability of graduating, meaning that the family of origin matters.

However, low upward mobility does not necessarily indicate a smaller opportunity to attain high levels of education. For example, upward mobility may be low in countries where a large share of parents has already attained tertiary education. For this reason, we introduce a second index, namely the loss of opportunity index. The next section is devoted to its definition.

#### 6.3.1.4 The opportunity index

In this section, we introduce the opportunity index and, to this end, we introduce the parameter $k$, which represents the probability for a child to have both parents without a tertiary degree. That is,

$$k = \Pr(\overline{PL})$$

and, consequently,

$$1 - k = 1 - \Pr\left(\overline{PL}\right) = \Pr(PL)$$

represents the probability that a child belongs to a family in which at least one parent has a tertiary education level.

We combine the information contained in $p$, $q$, and $k$ to define the opportunity index.

#### 6.3.2.1 The opportunity index

The probability that a child does not attain tertiary education independent of the educational level of the parents is given by

$$P(\overline{CL}) = kp + (1 - k)q.$$

**Definition:** The opportunity index, OI, is the complement to one of the probabilities that a child does not attain tertiary education:

$$\text{OI} = 1 - P(\overline{CL}) = 1 - kp + (1 - k)q.$$

We note that an opportunity index equal to zero (i.e., OI = 0) indicates the worst case, while an index equal to one (i.e., OI = 1) represents the best case.

#### 6.3.2.2 Ranking OECD countries by opportunity

With regard to the mobility index, MI, we rank OECD countries according to the opportunity index, OI. Table 6.2 reports these values. The minimum and maximum are 0.1425 and 0.6048, respectively. The average is 0.3956 with a standard error of 0.0248. Canada, Israel, and New Zealand have higher opportunity indices. At the bottom of the ranking are Turkey, Italy, and Austria.

### 6.3.3 Mobility vs. Opportunity

We now compare the two rankings. Figure 6.2 shows the comparison between the mobility index and the opportunity index. It is clear that the two rankings do not coincide. To measure the strength and direction of association between MI and OI, we perform the nonparametric Spearman rank-order correlation test. The value is 0.6342, that is, less than Pearson's correlation coefficient (i.e., 0.6653). The coefficient is positive but not very high and we conclude that, in general, large values of are sometimes associated with large values of OI. Moreover, Spearman's correlation coefficient is statistically significant because $p = 0.0005$, which is less than the usual threshold for the two-tailed test ($p < 0.05$). The value of Spearman's correlation coefficient suggests that

**Table 6.2** Ranking according to opportunity index.

| Rank | Country | Opportunity | Rank | Country | Opportunity |
|---|---|---|---|---|---|
| 1 | Canada | 0.6048 | 14 | Netherlands | 0.4027 |
| 2 | Israel | 0.5709 | 15 | Sweden | 0.3930 |
| 3 | New Zealand | 0.5638 | 16 | Spain | 0.3727 |
| 4 | Japan | 0.5331 | 17 | Germany | 0.3710 |
| 5 | Korea | 0.5283 | 18 | Poland | 0.3599 |
| 6 | Finland | 0.5183 | 19 | Chile | 0.3428 |
| 7 | Norway | 0.4761 | 20 | Slovenia | 0.3319 |
| 8 | Denmark | 0.4679 | 21 | Greece | 0.2954 |
| 9 | Estonia | 0.4522 | 22 | Czech Republic | 0.2309 |
| 10 | Australia | 0.4515 | 23 | Slovak Republic | 0.2306 |
| 11 | Ireland | 0.4375 | 24 | Austria | 0.2099 |
| 12 | United States | 0.4272 | 25 | Italy | 0.1661 |
| 13 | France | 0.4059 | 26 | Turkey | 0.1425 |

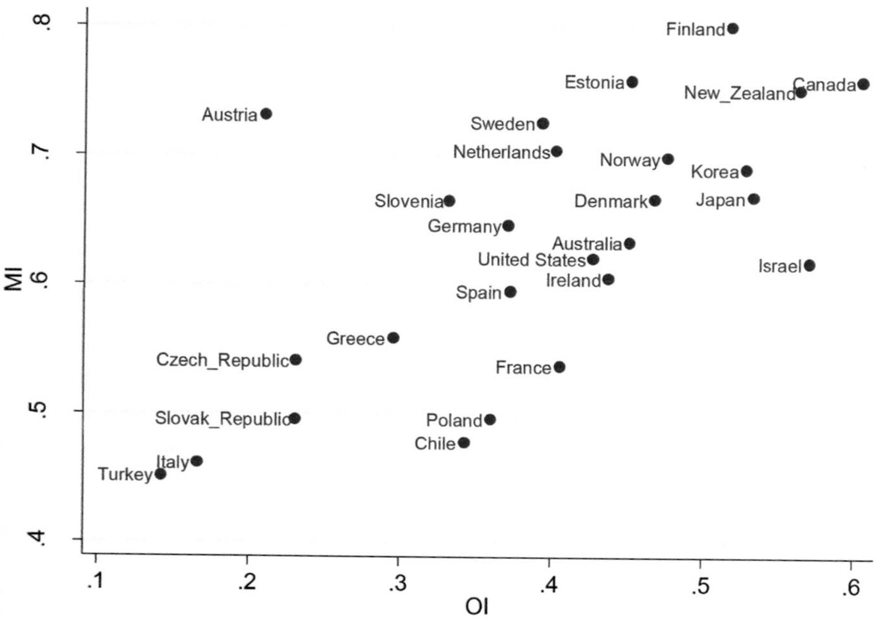


**Figure 6.2** Mobility vs opportunity.

there is not a perfect correlation between the two rankings and, as a result, MI and OI capture two different aspects of education.

When the rankings disagree, further criteria are needed to decide which country performs best. To clarify this point, we compare the results for two countries, namely Finland and New Zealand. According to MI, Finland is ranked at first and New Zealand is forth, with 0.8012 and 0.723, respectively. However, if we look at the ranking according to OI, New Zealand now occupies the third position and Finland the sixth, with 0.5638 and 0.5183, respectively. So, which of the two countries is the best? Keeping the two dimensions distinct does not help answer the question; so the two dimensions were combined. The next section is devoted to constructing a composite indicator that accounts for the two dimensions.

## 6.4 A New Aggregate for Mobility and Opportunity

As stressed in the previous section, when we compare the results based on MI and OI, we find conflicting results. We observe that countries with a high performance according to do not achieve similar results according to OI.

To overcome this situation, we follow Ciommi *et al.* (2017) and compute a composite index that combines the two dimensions (mobility and opportunity) using seven different aggregation methods.

We represent our data with a matrix $X$, whose entry $x_{ij}$ represents the value of the $j$th elementary indicator corresponding to the $i$th country, $j = 1, 2$, $i = 1, 2,\ldots,26$.

As proved by Ciommi *et al.* (2017), the following formula summarizes six methods for building the composite indicator for a given country $i$:

$$C_i = \left( \frac{1}{\sum_{j=1}^{2}(G_{.j})^{\alpha}} \cdot \sum_{j=1}^{2} r_{ij} \cdot \left(G_{.j}\right)^{\alpha} \right) - \beta \cdot (S_i \cdot cv_i)$$

where $\alpha = \{-1,0,1\}, \beta = \{0,1\}$, and

$$r_{ij} = \frac{x_{ij} - \underline{x}_j}{\overline{x}_{.j} - \underline{x}_j} \cdot 60 + 70$$

denotes the normalized indicators, obtained according to the Mazziotta and Pareto (2016) method (it covers the interval [70,130]), and $\underline{x}_j = \text{Rif}_{.j} - \left(\text{Max}_{.j} - \text{Min}_{.j}\right)/2$ and $\overline{x}_{.j} = \text{Rif}_{.j} + \left(\text{Max}_{.j} - \text{Min}_{.j}\right)/2$ are the two *goalposts* for MI and OI, respectively. $Max_{.j}$ and $Min_{.j}$ are, respectively, the maximum and minimum values of the two indicators (denoted by $j$) across all OECD countries, whereas $\text{Rif}_{.j}$ denotes the reference value, that is, the average value for any indicator. $G_{.j}$ represents the Gini index computed for indicator $j$ across all countries. Finally, $S_i$ and $cv_i$ denote, respectively, the standard deviation and coefficient of variation of all normalized indicators $r_{ij}$. The six indices can be determined by choosing combinations of $\alpha$ and $\beta$

The simplest index, EW, is obtained by fixing $\alpha = 0$ and $\beta = 0$. The resulting index is

$$C_i^{\text{EW}} = \frac{1}{2}\sum_{j=1}^{2} r_{ij}$$

which represents the simple mean. With the simple mean, we assume that the two indicators, namely MI and OI, have the same weight. In other words, we affirm that they have the same importance. This is a reasonable requirement when we have no information *a priori* about the relative importance of the variables defining the phenomenon.

The second method is the adjusted Mazziotta–Pareto method (Mazziotta and Pareto, 2013), hereafter AMPI. It is obtained from $C_i$ assuming $\alpha = 0$ and $\beta$Thus, for country $i$, we have

$$C_i^{\text{AMPI}} = \left(\frac{1}{2} \cdot \sum_{j=1}^{2} r_{ij}\right) - S_i \cdot cv_i.$$

As stressed by Mazziotta and Pareto, this method belongs to the non-compensatory composite measure, based on a penalty function $(S_i \cdot cv_i)$. The starting point for constructing the AMPI is the computation of the arithmetic mean of the elementary indicators, followed by an adjustment that accounts for horizontal variability of the indicators.

The third method, GW, aims to capture vertical inequality. It is a modification of the EW method where, instead of choosing the same weight for both indicators, we compute a weighted average of MI and OI with weights based on the Gini index of inequality. With this system of weights, we account for the distribution of indicators. Formally, we have

$$C_i^{\text{GW}} = \frac{1}{\sum_{j=1}^{2} G_{.j}} \cdot \sum_{j=1}^{2} r_{ij} \cdot G_{.j}.$$

GW weighs more unequal distributions more heavily (Chelli *et al.*, 2015); so if we believe that a more homogenous distribution of an indicator should imply a higher weight for that indicator, we use the inverse of the Gini index as weighting system. Thus, the definition of the iGW index is as follows:

$$C_i^{\text{iGW}} = \frac{1}{\sum_{j=1}^{2} (G_{.j})^{-1}} \cdot \sum_{j=1}^{2} r_{ij} \cdot (G_{.j})^{\alpha}.$$

The fifth and sixth methods are a combination of the GW, AMPI, iGW and GAMPI, respectively. In detail, for the former, the starting point is the GW, which we penalize as in the AMPI method. The resulting method is the so-called GAMPI, defined as

$$C_i^{\text{GAMPI}} = \left(\frac{1}{\sum_{j=1}^{2} G_{.j}} \cdot \sum_{j=1}^{2} r_{ij} \cdot G_{.j}\right) - S_i \cdot cv_i$$

whereas, for the former, iGAMPI, the starting point is the iGW, which we penalize according to the AMPI method. The resulting index is computed as follows:

$$C_i^{\mathrm{iGAMPI}} = \left( \frac{1}{\sum_{j=1}^{2} (G_{.j})^{-1}} \cdot \sum_{j=1}^{2} r_{ij} \cdot (G_{.j})^{-1} \right) - S_i \cdot cv_i.$$

Finally, the last method is a complete non-compensatory method, namely the geometric mean. It is defined as

$$C_i^{\mathrm{MGEO}} = \sqrt{\prod_{j=1}^{2} r_{ij}}.$$

The idea behind the use of a geometric mean is that if a country is at the minimum in one dimension, we do not want this component to be compensated by a high performance in the second dimension, as occurs for the arithmetic average.

### 6.4.1 A comparison of the methods

We compute composite indicators according to the seven proposed methods. Table 6.3 reports some basic descriptive statistics. The seven methods display very similar results. The penalization function does not affect the average value of the index very much, which decreases by about 0.5. has the lowest value, whereas has the maximum. The ranges are very similar, about 57 points. Moreover, all methods have negative asymmetry.

**Table 6.3** Descriptive statistics.

| | EW | AMPI | GW | iG | GAMPI | iGAMPI | GM |
|---|---|---|---|---|---|---|---|
| Mean | 100.00 | 99.53 | 100.00 | 100.00 | 99.53 | 99.53 | 99.77 |
| Standard error | 3.0289 | 3.0394 | 3.0260 | 3.0955 | 3.0471 | 3.0955 | 3.0330 |
| Median | 100.99 | 100.84 | 102.22 | 102.09 | 102.06 | 100.76 | 100.91 |
| Kurtosis | –0.5079 | –0.5989 | –0.4739 | –0.6377 | –0.5945 | –0.6747 | –0.5514 |
| Asymmetry | –0.4132 | –0.3574 | –0.4076 | –0.3923 | –0.3614 | –0.3444 | –0.3846 |
| Range | 56.43 | 56.40 | 57.59 | 56.21 | 57.56 | 55.86 | 56.42 |
| Min | 68.29 | 68.27 | 67.92 | 68.66 | 67.90 | 68.65 | 68.28 |
| Max | 124.73 | 124.68 | 125.51 | 124.87 | 125.47 | 124.50 | 124.70 |

**Table 6.4** Rankings.

| OECD | EW | AMPI | GW | iG | GAMPI | iGAMPI | GM |
|---|---|---|---|---|---|---|---|
| Australia | 12 | 12 | 11 | 13 | 10 | 12 | 12 |
| Austria | 17 | 18 | 19 | 12 | 22 | 16 | 18 |
| Canada | 1 | 1 | 1 | 2 | 1 | 2 | 1 |
| Chile | 22 | 22 | 22 | 23 | 21 | 23 | 22 |
| Czech_Republic | 23 | 23 | 23 | 22 | 23 | 21 | 23 |
| Denmark | 10 | 9 | 9 | 10 | 9 | 10 | 9 |
| Estonia | 4 | 4 | 7 | 4 | 7 | 4 | 4 |
| Finland | 2 | 2 | 3 | 1 | 3 | 1 | 2 |
| France | 19 | 19 | 18 | 19 | 18 | 19 | 19 |
| Germany | 15 | 15 | 15 | 15 | 15 | 13 | 15 |
| Greece | 20 | 20 | 21 | 20 | 20 | 20 | 20 |
| Ireland | 14 | 14 | 14 | 17 | 14 | 17 | 14 |
| Israel | 8 | 8 | 5 | 11 | 6 | 11 | 8 |
| Italy | 25 | 25 | 25 | 25 | 25 | 25 | 25 |
| Japan | 6 | 6 | 6 | 8 | 5 | 7 | 6 |
| Korea | 5 | 5 | 4 | 5 | 4 | 5 | 5 |
| Netherlands | 11 | 11 | 12 | 9 | 12 | 9 | 11 |
| New_Zealand | 3 | 3 | 2 | 3 | 2 | 3 | 3 |
| Norway | 7 | 7 | 8 | 6 | 8 | 6 | 7 |
| Poland | 21 | 21 | 20 | 21 | 19 | 22 | 21 |
| Slovak_Republic | 24 | 24 | 24 | 24 | 24 | 24 | 24 |
| Slovenia | 16 | 16 | 16 | 14 | 16 | 14 | 16 |
| Spain | 18 | 17 | 17 | 18 | 17 | 18 | 17 |
| Sweden | 9 | 10 | 10 | 7 | 11 | 8 | 10 |
| Turkey | 26 | 26 | 26 | 26 | 26 | 26 | 26 |
| United States | 13 | 13 | 13 | 16 | 13 | 15 | 13 |

Table 6.3 also reports the standard deviations for each method. The values are quite low, which means that the distribution of the composite indicators does not vary by much. Consequently, the values are very close. For this reason, it is interesting to compute the quintiles of the distribution for each method. Thus, for each country, we count the frequency of the ranking according to the quintile. Table 6.5 reports the results, where a darker colour corresponds to a higher frequency. As is clear, most of the rows present very

high values, meaning that the rank's quintile for a given country does not vary considerably. In detail, the countries analysed remain in the same quintile according to all methods 15 out of 26 times. For instance, Canada always falls in the first quintile, Denmark in the second, Germany the third, and so on.

**Table 6.5** Quintile rankings.

| **OECD** | *q1* | *q2* | *q3* | *q4* | *q5* |
|---|---|---|---|---|---|
| Australia | 0 | 2 | 5 | 0 | 0 |
| Austria | 0 | 0 | 2 | 4 | 1 |
| Canada | 7 | 0 | 0 | 0 | 0 |
| Chile | 0 | 0 | 0 | 1 | 6 |
| Czech_Republic | 0 | 0 | 0 | 1 | 6 |
| Denmark | 0 | 7 | 0 | 0 | 0 |
| Estonia | 5 | 2 | 0 | 0 | 0 |
| Finland | 7 | 0 | 0 | 0 | 0 |
| France | 0 | 0 | 0 | 7 | 0 |
| Germany | 0 | 0 | 7 | 0 | 0 |
| Greece | 0 | 0 | 0 | 7 | 0 |
| Ireland | 0 | 0 | 5 | 2 | 0 |
| Israel | 2 | 5 | 0 | 0 | 0 |
| Italy | 0 | 0 | 0 | 0 | 7 |
| Japan | 5 | 2 | 0 | 0 | 0 |
| Korea | 7 | 0 | 0 | 0 | 0 |
| Netherlands | 0 | 5 | 2 | 0 | 0 |
| New_Zealand | 7 | 0 | 0 | 0 | 0 |
| Norway | 2 | 5 | 0 | 0 | 0 |
| Poland | 0 | 0 | 0 | 6 | 1 |
| Slovak_Republic | 0 | 0 | 0 | 0 | 7 |
| Slovenia | 0 | 0 | 7 | 0 | 0 |
| Spain | 0 | 0 | 0 | 7 | 0 |
| Sweden | 0 | 7 | 0 | 0 | 0 |
| Turkey | 0 | 0 | 0 | 0 | 7 |
| United States | 0 | 0 | 7 | 0 | 0 |

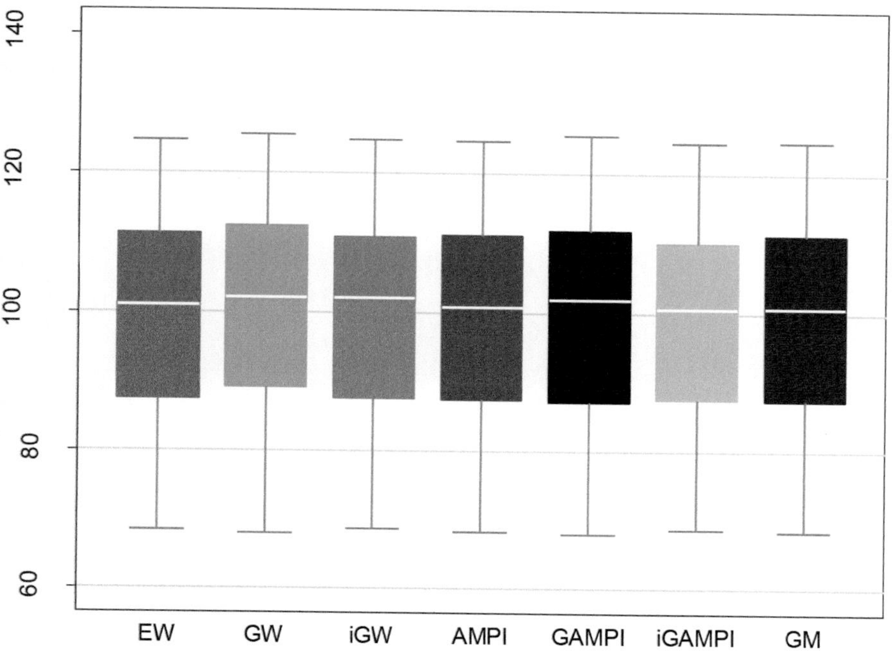


**Figure 6.3** Box plots.

Table 6.4 reports the ranking according to the seven methods. At the bottom of the ranking, Turkey and Italy occupy the last and penultimate positions, respectively, whereas at the top, Canada fluctuates between the first and second positions; it is ranked first in five out of seven methods. More specifically, it occupies the second position for the two methods involving the reciprocal of the Gini.

With regard to the values of the composite indicators again, Figure 6.3 presents box plots associated with the distributions for each synthetic method. The graph confirms that the methods exhibit similar results.

In sum, we find that the rankings are not reversed for a wide set of weighting vectors or aggregation schemes. Thus, according to Greco *et al.* (2019) and Cherchye *et al.* (2008), the proposed index shows a certain robustness.

Table 6.6 reports the quintiles for each indicator; so we can establish whether one or more methods rank differently. The results show that EW, AMPI, and GM display identical quintiles. When the quintiles differ across methods for a given country, GW and GAMPI show similar results in almost all countries. Counting the number of times there are differences among the rankings, we find that GAMPI differs the most (six times, compared with five times for iGAMPI, four for GW, and three for iG).

**Table 6.6** Quintile rankings.

| OECD | EW | AMPI | GW | iG | GAMPI | iGAMPI | GM |
|---|---|---|---|---|---|---|---|
| Australia | 3 | 3 | 2 | 3 | 2 | 3 | 3 |
| Austria | 4 | 4 | 4 | 3 | 5 | 3 | 4 |
| Canada | 1 | 1 | 1 | 1 | 1 | 1 | 1 |
| Chile | 5 | 5 | 5 | 5 | 4 | 5 | 5 |
| Czech Republic | 5 | 5 | 5 | 5 | 5 | 4 | 5 |
| Denmark | 2 | 2 | 2 | 2 | 2 | 2 | 2 |
| Estonia | 1 | 1 | 2 | 1 | 2 | 1 | 1 |
| Finland | 1 | 1 | 1 | 1 | 1 | 1 | 1 |
| France | 4 | 4 | 4 | 4 | 4 | 4 | 4 |
| Germany | 3 | 3 | 3 | 3 | 3 | 3 | 3 |
| Greece | 4 | 4 | 4 | 4 | 4 | 4 | 4 |
| Ireland | 3 | 3 | 3 | 4 | 3 | 4 | 3 |
| Israel | 2 | 2 | 1 | 2 | 1 | 2 | 2 |
| Italy | 5 | 5 | 5 | 5 | 5 | 5 | 5 |
| Japan | 1 | 1 | 1 | 2 | 1 | 2 | 1 |
| Korea | 1 | 1 | 1 | 1 | 1 | 1 | 1 |
| Netherlands | 2 | 2 | 3 | 2 | 3 | 2 | 2 |
| New Zealand | 1 | 1 | 1 | 1 | 1 | 1 | 1 |
| Norway | 2 | 2 | 2 | 1 | 2 | 1 | 2 |
| Poland | 4 | 4 | 4 | 4 | 4 | 5 | 4 |
| Slovak Republic | 5 | 5 | 5 | 5 | 5 | 5 | 5 |
| Slovenia | 3 | 3 | 3 | 3 | 3 | 3 | 3 |
| Spain | 4 | 4 | 4 | 4 | 4 | 4 | 4 |
| Sweden | 2 | 2 | 2 | 2 | 2 | 2 | 2 |
| Turkey | 5 | 5 | 5 | 5 | 5 | 5 | 5 |
| United States | 3 | 3 | 3 | 3 | 3 | 3 | 3 |

To confirm our intuition about preserving the ordering, we also compute country ranking correlations according to the seven methods. The results are reported in Figure 6.4 and Table 6.7. As is clear, the indices are highly correlated.

**Table 6.7 Spearman ranking correlation between methods.**

| | EW | GW | iGW | AMPI | GAMPI | iGAMPI | GM |
|---|---|---|---|---|---|---|---|
| **EW** | 1 | | | | | | |
| **GW** | 0.9884 | 1 | | | | | |
| **iGW** | 0.9747 | 0.9453 | 1 | | | | |
| **AMPI** | 0.9986 | 0.9904 | 0.9685 | 1 | | | |
| **GAMPI** | 0.9795 | 0.9945 | 0.9241 | 0.9843 | 1 | | |
| **iGAMPI** | 0.9843 | 0.9610 | 0.9911 | 0.9815 | 0.9494 | 1 | |
| **GM** | 0.9986 | 0.9904 | 0.9685 | 1 | 0.9843 | 0.9815 | 1 |

With respect to all correlation values, the index with the highest correlation with all other methods is EW[2], which implies that this may be used as the best aggregation method. More specifically, the correlation between EW and AMPI is 0.9986, which means that the two methods rank countries in almost the same way. Thus, from a computational point of view, EW and AMPI are the same. Since EW shows similar correlations with respect to all methods, we use the EW method.

In the last section, we investigate the relationship between the selected method and expenditures in education.

### 6.4.2 The relationship with expenditure

Finally, we test to see if there is any relationship between and the expenditure on educational institutions as a percentage of GDP per level of education.

We focus on the expenditure on tertiary education. We compute the correlation between the composite indicator constructed using the equal weight method and the expenditure on tertiary education (as a percentage of GDP). The value of the correlation is low, 0.4952, indicating that there is not a strict association between the two series.

The results are shown in Figure 6.5, from which some interesting conclusions can be drawn. First, as we know, Turkey and Italy have the lowest performance according to our composite index, but their percentages of expenditure on tertiary education are very different. In this case, therefore,

[2] We sum all correlations and find that correlates at 6.9241. The second ones are and, both with 6.9233. The method with the lowest correlation sum is, with 6.7722.

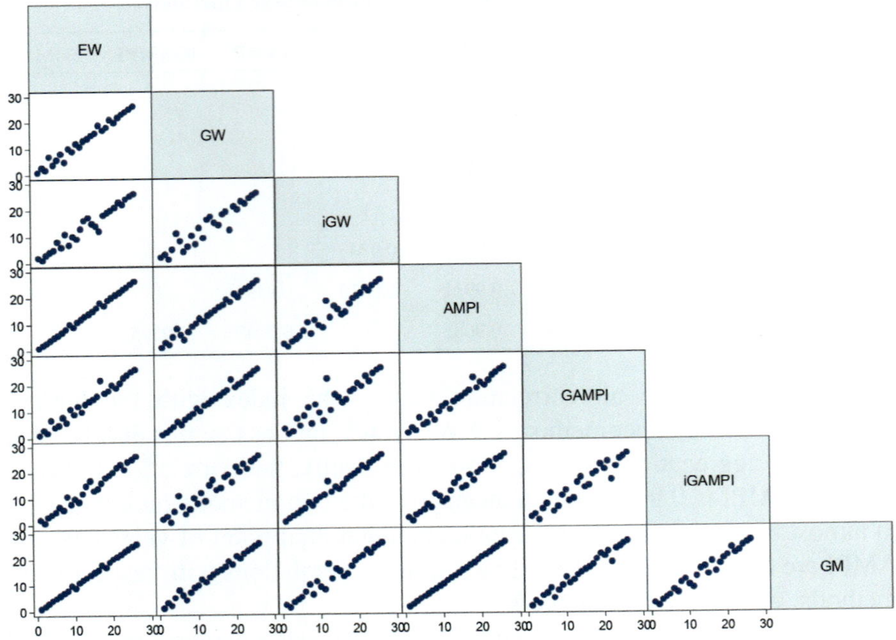


**Figure 6.4** Ranking correlations.

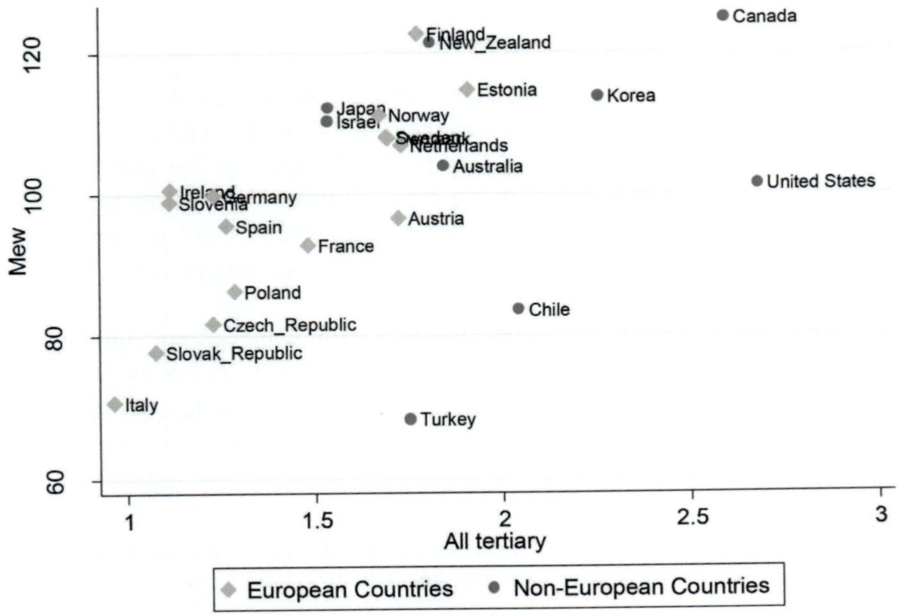


**Figure 6.5** EW versus expenditure.

the situation in Italy seems better than that in Turkey. Second, Turkey, Austria, and Finland seem to have the same percentage of expenditure, but the results based on the composite indicator are very different: Finland is ranked first, while Turkey occupies the last position. So it seems that it is not enough to increase the percentage of GDP devoted to education, especially for countries with low levels of GDP.

Figure 6.5 also compares European (diamonds; 17 countries) and non-European countries (circles; 9 countries). What is surprising is the difference in values for correlations in the two groups. For European countries, the correlation between expenditure and our index is very high (0.7915) compared to the correlation for non-European countries (0.1761). What emerges is that, for European countries, lower values of expenditure on tertiary education correspond to lower values in the composite indicator. As a consequence, the greater the percentage, the greater the value of the index. This is not true for non-European countries, however, as in the case of Australia and Turkey. Moreover, European countries display lower values both for our index and as to the percentage of GDP compared to non-European countries: 97.64987 versus 104.4391 for the index and 1.43 versus 2.01 for the percentage of GDP.

## 6.5 Conclusion

The aim of this paper was twofold. Driven by the idea that children can graduate more easily if they come from a family of graduates, we investigated two different aspects of the phenomenon, namely mobility and opportunity, using OECD data (OECD, 2017) for 26 OECD countries. On the other hand, since the two dimensions give contrasting results, we created a composite indicator and tested seven different aggregation methods. Looking at the different rankings, we found that the seven methods rank similarly and we concentrated on the method that correlates best with the other methods. With this specific dataset, we found that the equal weight method is the most appropriate and we propose it as the aggregate method to use. The first consideration is that, in this case, more sophisticated methods such as the Mazziotta–Pareto or methods, that is, modifications of the arithmetic mean but with weights based on the Gini index of inequality, give rankings that do not differ much with respect to the ranking computed according to the arithmetic mean (EW method). Since the arithmetic mean is the simplest aggregative method, we suggest using it as an aggregate method with the advantage that it is simple to calculate, especially for non-statisticians (i.e., policy makers or practitioners).

The index thus defined can be a useful instrument for very easily monitoring and evaluating the effectiveness of public policies. The main advantage of our proposal is that, in contrast to the mobility indices that consider only one dimension, here, we are able to capture two different pieces of information simultaneously: the difference in the probability of attaining tertiary education given the family status and the probability that a child does not attain tertiary education independent of the educational level attained by the parents. Thus, our proposal is more informative while preserving a simple functional form that requires an elementary computation.

Furthermore, intrigued by the international debate on the need to increase investment in education, we analysed the relationship between our index and the expenditure on tertiary education as percentage of GDP. What emerges is that two groups exist. European and non-European countries display very different behaviour. While a good correlation emerges for European countries (the correlation coefficient is about 0.8), it seems almost non-existent for non-European countries (coefficient less than 0.2).

This suggests the need of further investigations. First, it could be of potential interest to look at per-capita expenditure on tertiary education instead of the expenditure as a percentage of GDP. Second, from a theoretical point of view, we would like to propose a parametric aggregation method where the choice of parameter could be a function that reflects several aspects of well-being, for instance, by maximizing the proportion of explained variance in the model.

## References

Bleys, B. (2012). Beyond GDP: Classifying alternative measures for progress. Social Indicators Research, 109(3), 355-376.

Chelli, F. M., Ciommi, M., Emili, A., Gigliarano, C., Taralli, S. (2015). Comparing equitable and sustainable well-being (BES) across the Italian provinces. A factor analysis-based approach. Rivista Italiana di Economia Demografia e Statistica, 69(3), 61-72.

Cherchye, L., Ooghe, E., Van Puyenbroeck, T. (2008). Robust human development rankings. Journal of Economic Inequality, 6(4), 287–321.

Ciommi, M., Gigliarano, C., Emili, A., Taralli, S., Chelli, F. M. (2017). A new class of composite indicators for measuring well-being at the local level: An application to the equitable and sustainable well-being (BES) of the Italian provinces. Ecological Indicators, 76, 281-296.

Conlisk, J., Sommers, P. (1979). Eigenvector status proxies in Markov chain mobility models. Sociological Methods & Research, 8(2), 159-178.

Greco, S., Ishizaka, A., Tasiou, M., Torrisi, G. (2019). On the methodological framework of composite indices: A review of the issues of weighting, aggregation, and robustness. Social Indicators Research, 141(1), 61-94.

Mazziotta, M., Pareto, A. (2013). Methods for constructing composite indices: One for all or all for one. Rivista Italiana di Economia Demografia e Statistica, 67(2), 67-80.

Mazziotta, M., Pareto, A. (2016). On a generalized non-compensatory composite index for measuring socio-economic phenomena. Social Indicators Research, 127(3), 983-1003.

OECD. (2008). *Handbook on Constructing Composite Indicators. Methodology and User Guide*. OECD Publications, Paris. https://www.oecd.org/els/soc/handbookonconstructingcompositeindicatorsmethodologyanduserguide.htm

OECD. (2017). *Education at a Glance 2017: OECD Indicators*. OECD Publishing, Paris. http://dx.doi.org/10.1787/eag-2017-en.

Shorrocks, A. F. (1978). The measurement of mobility. Econometrica: Journal of the Econometric Society, 1013-1024.

Van de Gaer, D., Schokkaert, E., Martinez, M. (2001). Three meanings of intergenerational mobility. Economica, 68(272), 519-538.

# Index

**S**



**T**



**U**

# About the Editor

**Margherita Carlucci** is Full Professor of Economic Statistics at the Department of Social and Economics Sciences of Sapienza University of Rome, member of Sapienza University Quality Assurance Committee. Former Director of the Department of Economics of Sapienza University of Rome and Vice-President of IS.CO.NA (Institute of National Accounts), currently her research interests include integrated systems of economic, social and environmental accounting; the relationship between socioeconomic factors and land degradation; the dynamics of urban development; the definition and measurement of indicators of sustainable development.

For Product Safety Concerns and Information please contact our EU representative GPSR@taylorandfrancis.com Taylor & Francis Verlag GmbH, Kaufingerstraße 24, 80331 München, Germany